Solutions Manual for Beginning Partial Differential Equations

PURE AND APPLIED MATHEMATICS

A Wiley Series of Texts, Monographs, and Tracts

Founded by RICHARD COURANT
Editors Emeriti: MYRON B. ALLEN III, PETER HILTON, HARRY HOCHSTADT, ERWIN KREYSZIG, PETER LAX, JOHN TOLAND

A complete list of the titles in this series appears at the end of this volume.

Solutions Manual for Beginning Partial Differential Equations

Third Edition

Peter V. O'Neil
University of Alabama at Birmingham

WILEY

Published by John Wiley & Sons, Inc., Hoboken, New Jersey

Published simultaneously in Canada

Library of Congress Cataloging in Publication Data is available.

ISBN 978-1-118-63009-9

Printed in the United States of America
10 9 8 7 6 5 4 3 2 1

Contents

Preface

This manual contains solutions for many of the problems in Beginning Partial Differential Equations, third edition.

Because solutions for many odd-numbered problems are included in Chapter Nine of the book, most of the problems included here are even-numbered. However, particularly in the case of problems exploring ideas beyond the text discussion, some odd-numbered solutions are also included.

Chapter 1

First Ideas

1.1 Two Partial Differential Equations

2. Verifying that the function is a solution of the heat equation is a straightforward exercise in differentiation. One way to show that $u(x,t)$ is unbounded is to observe that if $t > 0$ and $x = 2\sqrt{kt}$, then

$$u(x,t) = \frac{1}{e}t^{-3/2}$$

and this can be made as large as we like by choosing t sufficiently close to zero.

4. By the chain rule,

$$\begin{aligned} u_x &= \frac{1}{2}(f'(x-ct) + f'(x+ct)), \\ u_{xx} &= \frac{1}{2}(f''(x-ct) + f''(x+ct)), \\ u_t &= \frac{1}{2}(f'(x-ct)(-c) + f'(x+ct)(c)), \text{ and} \\ u_{tt} &= \frac{1}{2}(f''(x-ct)(-c)^2 + f''(x+ct)(c)^2). \end{aligned}$$

It is routine to verify that $u_{tt} = c^2 u_{xx}$.

7. One way to show that the transformation is one to one is to evaluate the Jacobian

$$\begin{vmatrix} \xi_x & \xi_t \\ \eta_x & \eta_t \end{vmatrix} = \begin{vmatrix} 1 & a \\ 1 & b \end{vmatrix} = b - a \neq 0.$$

Solutions Manual to Accompany Beginning Partial Differential Equations,
Third Edition. Peter V. O'Neil.

Finally, solve $\xi = a + at$, $\eta = x + bt$ for x and t to obtain the inverse transformation

$$x = \frac{1}{b-a}(b\xi - a\eta), t = \frac{1}{b-a}(\eta - \xi).$$

8. With $V(\xi,\eta) = u(x(\xi,\eta), t(\xi,\eta))$, chain rule differentiations yield:

$$\begin{aligned} u_x &= V_\xi \xi_x + V_\eta \eta_x = V_\xi + V_\eta, \\ u_t &= V_\xi \xi_t + V_\eta \eta_t = aV_\xi + bV_\eta, \end{aligned}$$

and, by continuing these chain rule differentiations and using the product rule,

$$\begin{aligned} u_{xx} &= V_{\xi\xi} + 2V_{\xi\eta} + V_{\eta\eta}, \\ u_{tt} &= a^2 V_{\xi\xi} + 2abV_{\xi\eta} + b^2 V_{\eta\eta}, \text{ and} \\ u_{xt} &= aV_{\xi\xi} + (a+b)V_{\xi\eta} + bV_{\eta\eta}. \end{aligned}$$

Now collect terms to obtain

$$\begin{aligned} &Au_{xx} + Bu_{xt} + Cu_{tt} = \\ &\quad (A + aB + a^2C)V_{\xi\xi} + (2A + (a+b)B + 2abC)V_{\xi\eta} + (A + bB + b^2C)V_{\eta\eta}. \end{aligned}$$

This, coupled with the fact that $H(x,t,u,u_x,u_t)$ transforms to some function $K(\xi,\eta,V,V_\xi,V_\eta)$, yields the conclusion.

9. From the solution of problem 8, the transformed equation is hyperbolic if $C \neq 0$ because in that case we can choose a and b to make the coefficients of $V_{\xi\xi}$ and $V_{\eta\eta}$ vanish. This is done by choosing a and b to be the distinct roots of

$$A + Ba + Ca^2 = 0 \text{ and } A + Bb + Cb^2$$

which are the same quadratic equation. For example, we could choose

$$a = \frac{-B + \sqrt{B^2 - 4AC}}{2C} \text{ and } b = \frac{-B - \sqrt{B^2 - 4AC}}{2C}.$$

If $C = 0$, use the transformation

$$\xi = t,\ \eta = -\frac{B}{A}x + t.$$

Now chain rule differentiations yield

$$\begin{aligned} u_x &= -\frac{B}{A}V_\eta,\ u_t = V_\xi + V_\eta, \\ u_{xx} &= \frac{B^2}{A^2}V_{\eta\eta},\ u_{xt} = -\frac{B}{A}V_{\xi\eta} - \frac{B}{A}V_{\eta\eta}. \end{aligned}$$

We do not need u_{tt}, because $C = 0$ in this case. Now we obtain

$$Au_{xx} + Bu_{xt} + Cu_{tt} = -\frac{B^2}{A}V_{\xi\eta},$$

yielding a hyperbolic canonical form

$$V_{\xi\eta} + K(\xi, \eta, V, V_\xi, V_\eta) = 0$$

of the given partial differential equation.

10. In this case suppose $B^2 - 4AC = 0$. Now let

$$\xi = x,\ \eta = x - \frac{B}{2C}t.$$

Now

$$\begin{aligned} u_x &= V_\xi + V_\eta,\ u_t = -\frac{B}{2C}V_\eta, \\ u_{xx} &= V_{\xi\xi} + 2V_{\xi\eta} + V_{\eta\eta},\ u_{tt} = \frac{B^2}{4C^2}V_{\eta\eta}, \text{ and} \\ u_{xt} &= -\frac{B}{2C}V_{\xi\eta} - \frac{B}{2C}V_{\eta\eta}. \end{aligned}$$

Then

$$\begin{aligned} &Au_{xx} + Bu_{xt} + Cu_{tt} \\ &= A(V_{\xi\xi} + 2V_{\xi\eta} + V_{\eta\eta}) - \frac{B^2}{2C}(V_{\xi\eta} + V_{\eta\eta}) + \frac{B^2}{4C}V_{\eta\eta} \\ &= AV_{\xi\xi} + V_{\xi\eta}\left(2A - \frac{B^2}{2C}\right) + V_{\eta\eta}\left(A - \frac{B^2}{2C} + \frac{B^2}{4C}\right) \\ &= AV_{\xi\xi}, \end{aligned}$$

with two terms on the next to last line vanishing because $B^2 - 4AC = 0$. This gives the canonical form

$$V_{\xi\xi} + K(\xi, \eta, V, V_\xi, V_\eta) = 0$$

for the original partial differential equation when $B^2 - 4AC = 0$.

11. Suppose now that $B^2 - 4AC < 0$. Let the roots of $Ca^2 + Ba + A = 0$ be $p \pm iq$. Let

$$\xi = x + pt,\ \eta = qt.$$

Proceeding as in the preceding two problems, we find that

$$\begin{aligned} &Au_{xx} + Bu_{xt} + Cu_{tt} \\ &= (A + Bp + Cp^2)V_{\xi\xi} + (qB + 2pqC)V_{\xi\eta} + q^2V_{\eta\eta}. \end{aligned}$$

Now we need some information about p and q. Because of the way $p+iq$ was chosen,

$$C(p+iq)^2 + B(p+iq) + A = 0.$$

This gives us

$$Cp^2 - Cq^2 + Bp + A + (2Cpq + Bq)i = 0.$$

Then

$$Cp^2 - Cq^2 + Bp = 0 \text{ and } 2Cpq + Bq = 0.$$

In this case,

$$Au_{xx} + Bu_{xt} + Cu_{tt} = q^2(V_{\xi\xi} + V_{\eta\eta})$$

and we obtain the canonical form

$$V_{\xi\xi} + V_{\eta\eta} + K(\xi, \eta, V, V_\xi, V_\eta) = 0$$

for this case.

12. The diffusion equation is parabolic and the wave equation is hyperbolic.

14. $B^2 - 4AC = 33 > 0$, so the equation is hyperbolic. With

$$a = \frac{1+\sqrt{33}}{8} \text{ and } b = \frac{1-\sqrt{33}}{8}$$

the canonical form is

$$V_{\xi\eta} - \frac{16}{49\sqrt{33}}\left(\frac{-7-\sqrt{33}}{8}\xi + \frac{7-\sqrt{33}}{8}\eta\right).$$

16. With $A = 1, B = 0$, and $C = 0$, $B^2 - 4AC = -36 < 9$, so the equation is elliptic. Solve $9a^2 + 1 = 0$ to get $a = \pm i/3$. Thus use the transformation

$$\xi = x,\ \eta = \frac{1}{3}t$$

to obtain the canonical form

$$V_{\xi\xi} + V_{\eta\eta} + \xi^2 - 3\eta V = 0.$$

1.2 Fourier Series

2. $\cos(3x)$ is the Fourier series of $\cos(3x)$ on $[-\pi, \pi]$. This converges to $\cos(3x)$ for $-\pi \leq x \leq \pi$.

4. The Fourier series of $f(x)$ on $[-2, 2]$ is

$$\sum_{n=1}^{\infty} \frac{4(1-(-1)^n)}{n^2\pi^2}\cos(n\pi x/2),$$

converging to $1 - |x|$ for $-2 \leq x \leq 2$. Figure 1.1 compares a graph of $f(x)$ with the fifth partial sum of the series.

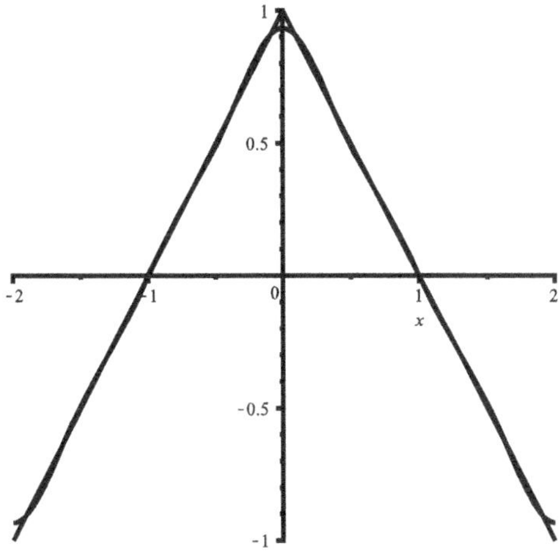

Figure 1.1: $f(x)$ and the 5th partial sum of the Fourier series in Problem 4.

6. The Fourier series is

$$\frac{2}{\pi} + \frac{4}{3\pi}\cos(x) - \sin(x)$$
$$+ \sum_{n=2}^{\infty} \frac{4(-1)^{n+1}}{\pi(4n^2-1)}\cos(nx).$$

Figure 1.2 compares a graph of the function with the fifth partial sum of the series.

8. The Fourier series converges to

$$\begin{cases} \cos(x) & \text{for } -2 < x < 1/2, \\ \sin(x) & \text{for } 1/2 < x < 2, \\ (\cos(2)+\sin(2))/2 & \text{for } x = \pm 2. \end{cases}$$

10. The series converges to

$$\begin{cases} 1 & \text{for } -2 < x < 0, \\ -1 & \text{for } 0 < x, 1/2, \\ x^2 & \text{for } 1/2 < x < 2, \\ 0 & \text{at } x = 0, \\ -3/8 & \text{at } x = 1/2, \\ 5/2 & \text{at } x = \pm 2. \end{cases}$$

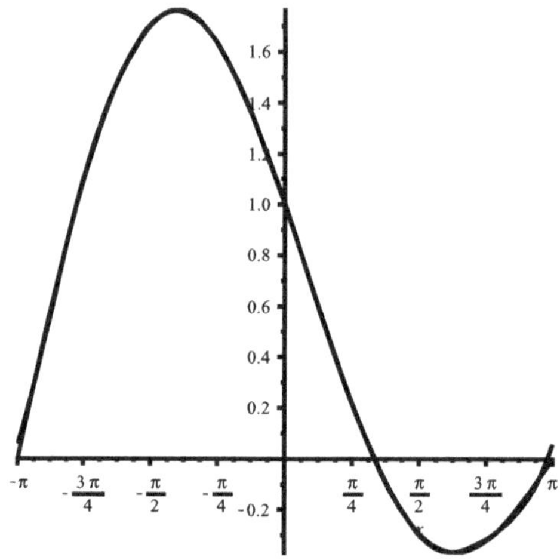

Figure 1.2: $f(x)$ and the 5th partial sum of the Fourier series in Problem 6.

12. The series converges to

$$\begin{cases} 1-x & \text{for } -3 < x < -1/2, \\ 2+x & \text{for } -1/2 < x < 1, \\ 4-x^2 & \text{for } 1 < x < 2, \\ 1-x-x^2 & \text{for } 2 < x < 3, \\ 3/2 & \text{at } x = -1/2, \\ 3 & \text{at } x = 1, \\ -5/2 & \text{at } x = 2, \\ -7/2 & \text{at } x = \pm 3. \end{cases}$$

14. Multiply by $f(x)$ to obtain

$$\begin{aligned}(f(x))^2 =& \frac{1}{2}a_0 f(x) \\ &+ \sum_{n=1}^{\infty} \left(a_n f(x)\cos(n\pi x/L) + b_n f(x)\sin(n\pi x/L)\right).\end{aligned}$$

Integrate term by term:

$$\begin{aligned}&\int_{-L}^{L} (f(x))^2\,dx = \frac{1}{2}a_0 \int_{-L}^{L} f(x)\,dx \\ &+ \sum_{n=1}^{\infty} \left(a_n \int_{-L}^{L} f(x)\cos(n\pi x/L)\,dx + b_n \int_{-L}^{L} f(x)\sin(n\pi x/L)\,dx\right).\end{aligned}$$

Then

$$\int_{-L}^{L} (f(x))^2\,dx = \frac{1}{2}a_0(La_0) + \sum_{n=1}^{\infty} L(a_n^2 + b_n^2).$$

Upon division by L, this yields Parseval's equation.

16. The cosine series is

$$\sum_{n=1}^{\infty} \frac{4\sin(n\pi/2)}{n\pi}\cos(n\pi x/2),$$

converging to 1 for $0 \le x < 1$, to -1 for $1 < x \le 2$, and to 0 at $x = 1$. Figure 1.3 compares the function to the 100th partial sum of this cosine expansion.

The sine series is

$$\sum_{n=1}^{\infty} \frac{1}{n\pi}(-4\cos(n\pi/2) + 2(1 + (-1)^n))\sin(n\pi x/2),$$

converging to 0 at the end points and at 1, and to the function for $0 < x < 1$ and $1 < x < 2$. Figure 1.4 is the 100th partial sum of this sine series.

18. The cosine expansion is

$$1 + \sum_{n=1}^{\infty} \frac{4}{n^2\pi^2}(-1 + (-1)^n)\cos(n\pi x).$$

This converges to $f(x)$ on $[0, 1]$. Figure 1.5 compares the function with the 10th partial sum of this cosine series.

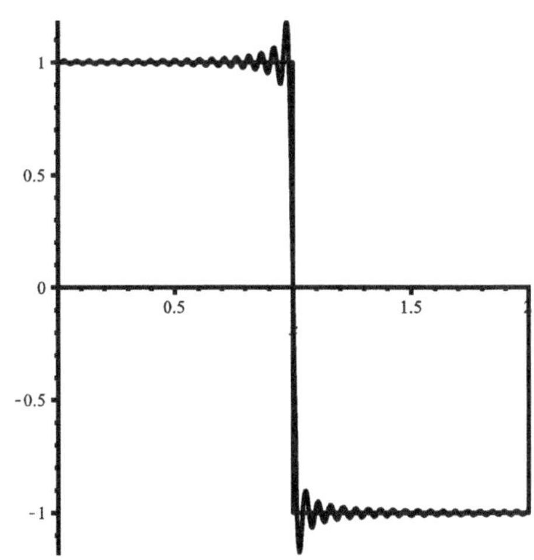

Figure 1.3: $f(x)$ and the 100th partial sum of the cosine series in Problem 16.

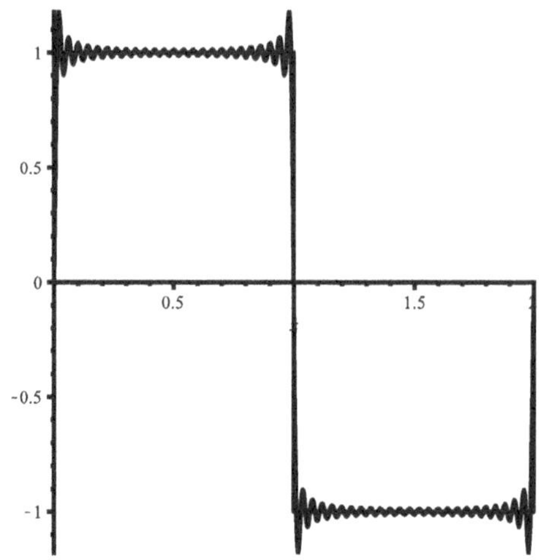

Figure 1.4: $f(x)$ and the 100th partial sum of the sine expansion in Problem 16.

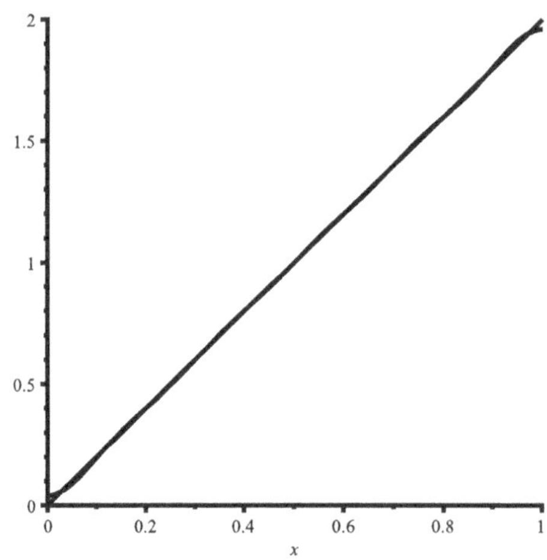

Figure 1.5: $f(x)$ and the 10th partial sum of the cosine series in Problem 18.

The sine expansion is

$$\sum_{n=1}^{\infty} \frac{4}{n\pi}(-1)^{n+1} \sin(n\pi x),$$

converging to 0 at $x = 0$ and $x = 1$, and to $2x$ for $0 < x < 1$. Figure 1.6 compares the function with the 50th partial sum of this sine expansion.

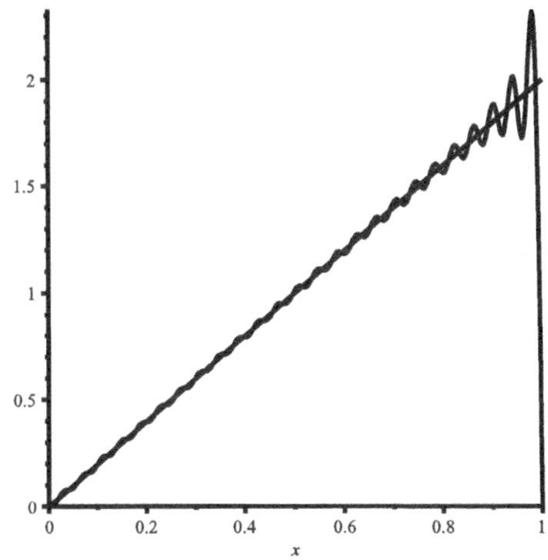

Figure 1.6: $f(x)$ and the 50th partial sum of the sine expansion in Problem 18.

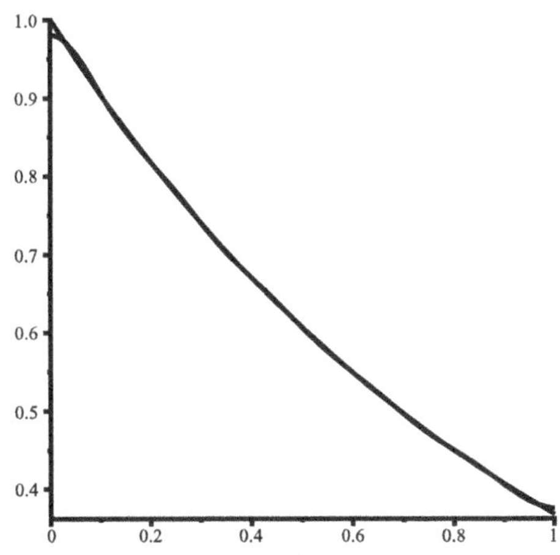

Figure 1.7: $f(x)$ and the 10th partial sum of the cosine series in Problem 20.

20. The cosine expansion is

$$1 - \frac{1}{e} + \sum_{n=1}^{\infty} \frac{2}{1 + n^2\pi^2}(1 - e^{-1}(-1)^n)\cos(n\pi x),$$

converging to e^{-x} for $0 \leq x \leq 1$. Figure 1.7 shows the function and the 10th partial sum of this series.

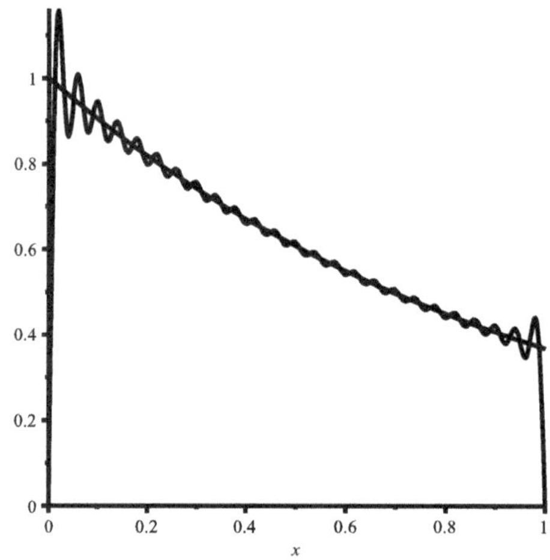

Figure 1.8: $f(x)$ and the 50th partial sum of the sine expansion in Problem 20.

The sine expansion is

$$\sum_{n=1}^{\infty} \frac{2n\pi}{1+n^2\pi^2}(1-e^{-1}(-1)^n)\sin(n\pi x).$$

This series converges to 0 at $x=0$ and at $x=1$, and to e^{-x} for $0<x<1$. Figure 1.8 shows the 50th partial sum.

22. The cosine expansion is

$$\frac{1}{2}+\sum_{n=1}^{\infty} \frac{4}{n^2\pi^2}(2\cos(n\pi/2)-(1+(-1)^n))\cos(n\pi x/2),$$

converging to $f(x)$ on $[0,2]$. Figure 1.9 shows graphs of the function and the 10th partial sum of this cosine series.

The sine series is

$$\sum_{n=1}^{\infty} \frac{16\sin(n\pi x/2)}{n^2\pi^2}\sin(n\pi x/2),$$

converging to $f(x)$ on $[0,2]$. The function and the 10th partial sum of this sine series are shown in Figure 1.10.

23. Expand $f(x)=\sin(x)$ in a cosine series on $[0,\pi]$:

$$\sin(x)=\frac{2}{\pi}+\sum_{n=2}^{\infty} \frac{-2(1+(-1)^n)}{\pi(n^2-1)}\cos(nx).$$

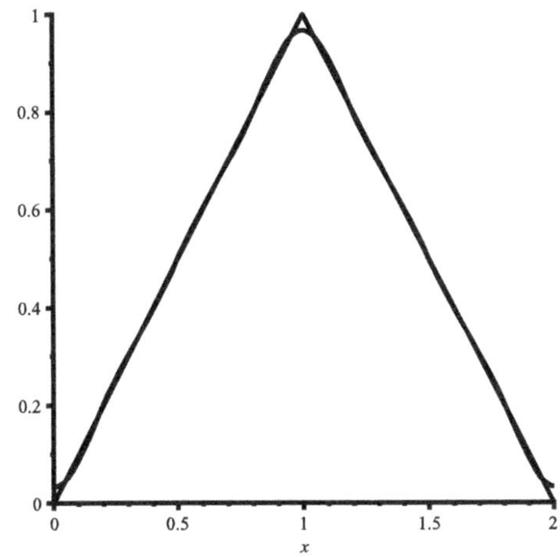

Figure 1.9: $f(x)$ and the 10th partial sum of the cosine series in Problem 22.

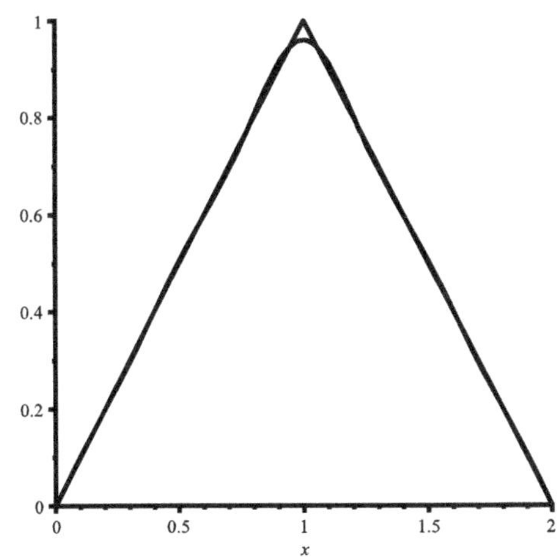

Figure 1.10: $f(x)$ and the 50th partial sum of the sine expansion in Problem 22.

Since $1 + (-1)^n = 0$ if n is odd, we need only to retain the even positive integers in the sum. Replace n with $2n$ to write

$$\sin(x) = \sum_{n=1}^{\infty} \frac{-4}{\pi(4n^2 - 1)} \cos(2nx).$$

Now choose $x = \pi/2$.

1.3 Two Eigenvalue Problems

2. Eigenvalues and eigenfunctions are

$$\lambda_n = \left(\frac{(2n-1)\pi}{2L}\right)^2, \; X_n(x) = \cos\left(\frac{(2n-1)\pi x}{2L}\right).$$

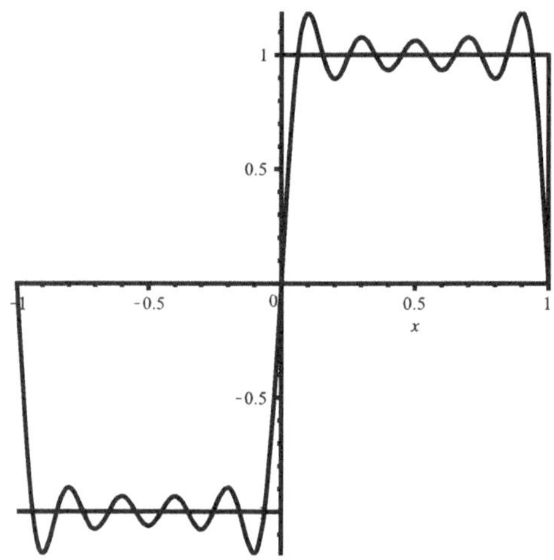

Figure 1.11: $f(x)$ and the 10th partial sum.

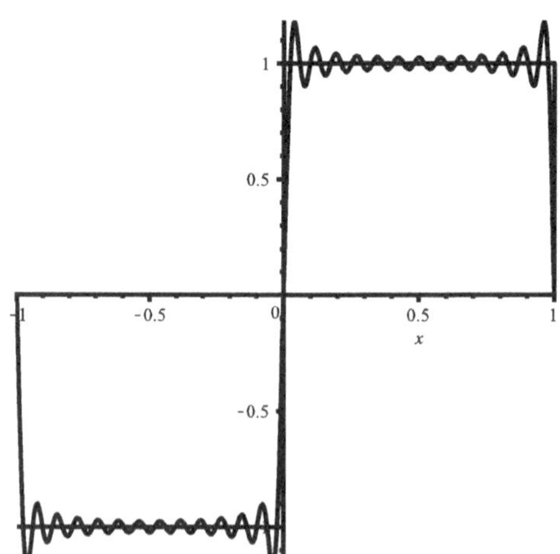

Figure 1.12: $f(x)$ and the 25th partial sum.

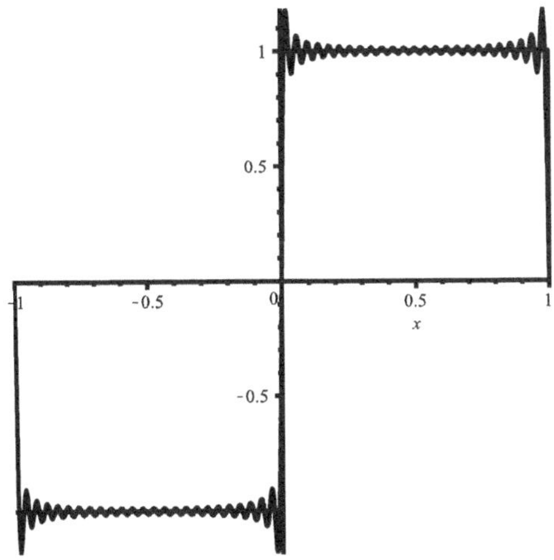

Figure 1.13: $f(x)$ and the 50th partial sum.

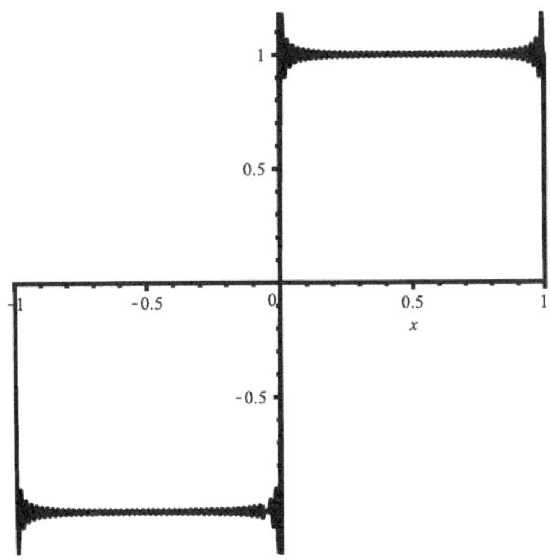

Figure 1.14: $f(x)$ and the 100th partial sum.

4. Eigenvalues and eigenfunctions are

$$\lambda_n = \alpha_n^2,\ X_n(x) = \sin(\alpha_n x),$$

where α_n is the nth positive root (in increasing order) of the equation $\tan(\alpha L) = -2\alpha$.

1.4 A Proof of the Convergence Theorem

The Fourier series of $f(x)$ on $[-1, 1]$ is

$$\sum_{n=1}^{\infty} \frac{2}{n\pi}(1 - (-1)^n) \sin(n\pi x).$$

Figures 1.11–1.14 show the function and the nth partial sum for $n = 10, 25, 50, 100$, respectively.

Chapter 2

Solutions of the Heat Equation

2.1 Solutions on an Interval $[0, L]$

2. By inspection the solution is $u(x,t) = T$.

4. By equation 2.2 the solution is

$$u(x,t) = \sin(\pi x)e^{-k\pi^2 t}.$$

6. Let $u(x,t) = U(x,t) + \psi(x)$. For U to satisfy the standard heat equation choose $\psi(x)$ so that $\psi''(x) = 0$. For homogeneous boundary conditions on the problem for $U(x,t)$, we also want $\psi(0) = 3$ and $\psi(5) = \sqrt{7}$. These conditions determine $\psi(x)$:

$$\psi(x) = \frac{\sqrt{7}-3}{5}x + 3.$$

The solution of the problem is

$$u(x,t) = U(x,t) + \frac{\sqrt{7}-3}{5}x + 3,$$

where

$$U(x,t) = \sum_{n=1}^{\infty} b_n \sin(n\pi x/5)e^{-kn^2\pi^2 t/25}$$

Solutions Manual to Accompany Beginning Partial Differential Equations, Third Edition. Peter V. O'Neil.

and

$$\begin{aligned} b_n &= \frac{2}{5}\int_0^5 (x^2 - \psi(x))\sin(n\pi x/5)\,dx \\ &= \frac{2}{n^3\pi^3}\left((-3-(25-\sqrt{7})(-1)^n)n^2\pi^2 + 50((-1)^n - 1)\right). \end{aligned}$$

8. Let $u(x,t) = U(x,t) + \psi(x)$. To obtain a standard problem for U, choose

$$\psi(x) = \frac{3}{5}x + 1.$$

The problem for $U(x,t)$ has the solution

$$U(x,t) = \sum_{n=1}^{\infty} b_n \sin(n\pi x/5)e^{-7n^2\pi^2 t/25},$$

where

$$\begin{aligned} b_n &= \frac{2}{5}\int_0^5 (e^{-x} - \psi(x))\sin(n\pi x/5)\,dx \\ &= \frac{1}{n\pi(25+n^2\pi^2)}\left(2n^2\pi^2(-1)^{n+1} + 200(-1)^n + 8n^2\pi^2(-1)^n - 50\right). \end{aligned}$$

10. Let $u(x,t) = e^{\alpha x + \beta t}v(x,t)$ in the partial differential equation to obtain

$$v_t = (k\alpha^2 - h\alpha - \beta)v + (2k\alpha - h)v_x + kv_{xx}.$$

Simplify this equation by choosing α and β to make the coefficients of v and v_x equal to zero. Thus choose

$$\alpha = \frac{h}{2k},\ \beta = -\frac{h^2}{4k}.$$

The problem for $v(x,t)$ is

$$\begin{aligned} v_t &= kv_{xx}, \\ v(0,t) &= v(L,t) = 0, \\ v(x,0) &= e^{-hx/2k}f(x). \end{aligned}$$

This problem has the solution

$$v(x,t) = \sum_{n=1}^{\infty} b_n \sin(n\pi x/L)e^{-kn^2\pi^2 t/L^2},$$

where

$$b_n = \frac{2}{L}\int_0^L e^{-3\xi}f(\xi)\sin(n\pi\xi/L)\,d\xi.$$

12. Let $k = 1$ and $h = 6$ in problem 10 to set

$$u(x,t) = e^{3x-6t}v(x,t).$$

The problem for v is

$$\begin{aligned} v_t &= u_{xx}, \\ v(0,t) &= v(\pi,t) = 0, \\ v(x,0) &= e^{-3x}f(x). \end{aligned}$$

The solution for v is

$$v(x,t) = \sum_{n=1}^{\infty} b_n \sin(nx)e^{-n^2 t},$$

where

$$\begin{aligned} b_n &= \frac{2}{\pi}\int_0^{\pi} e^{-3\xi}\sin(\xi)\sin(n\xi)\,d\xi \\ &= \frac{12n}{n^4+16n^2+100}\left(1+(-1)^n e^{-3\pi}\right). \end{aligned}$$

14. Let $u(x,t) = e^{-8t}v(x,t)$. The problem for $v(x,t)$ is

$$\begin{aligned} v_t &= kv_{xx}, \\ v_x(0,t) &= v_x(2\pi,t) = 0, \\ v(x,0) &= x(2\pi - x). \end{aligned}$$

This has the solution

$$v(x,t) = \frac{2}{3}\pi^2 - \sum_{n=1}^{\infty}\frac{4}{n^2}(1+(-1)^n)\cos(nx/2)e^{-kn^2t/4}.$$

The solution of the original problem is $u(x,t) = e^{-8t}v(x,t)$.

16. By inspection, the solution is $u(x,t) = B$.

18. The problem is

$$\begin{aligned} u_t &= ku_{xx}, \\ u(0,t) &= u_x(L,t) = 0, \\ u(x,0) &= B. \end{aligned}$$

Upon letting $u(x,t) = X(x)T(t)$, we obtain

$$\begin{aligned} &X'' + \lambda X = 0,\ X(0) = X'(L) = 0 \\ &T' + k\lambda X = 0. \end{aligned}$$

The eigenvalues and eigenfunctions are

$$\lambda_n = \left(\frac{(2n-1)\pi}{2L}\right)^2, X_n(x) = \sin((2n-1)\pi x/2L).$$

We also obtain

$$T_n(t) = e^{-(2n-1)^2\pi^2 kt/4L^2}.$$

Thus try a solution

$$u(x,t) = \sum_{n=1}^{\infty} b_n \sin((2n-1)\pi x/2L) e^{-(2n-1)^2\pi^2 kt/4L^2}.$$

To find the coefficients, we need

$$u(x,0) = B = \sum_{n=1}^{\infty} b_n \sin((2n-1)\pi x/2L).$$

This is not a Fourier series. However, observe that

$$\int_0^L X_n(x)X_m(x)\,dx = \begin{cases} 0 & \text{if } n \neq m, \\ L/2 & \text{if } n = m. \end{cases}$$

Using the same informal reasoning used to derive the Fourier coefficients, multiply the series for $u(x,0)$ by $\sin((2m-1)\pi x/2l)$ and integrate term by term to obtain

$$b_n = \frac{2}{L}\int_0 B \sin((2n-1)\pi x/2L)\,dx = \frac{2B}{L}.$$

The solution is

$$u(x,t) = \frac{2B}{L}\sum_{n=1}^{\infty} \sin((2n-1)\pi x/2L) e^{-(2n-1)^2\pi^2 kt/4L^2}.$$

20. The problem for the temperature distribution function is

$$\begin{aligned} u_t = ku_{xx} &= -A(u-T), \\ u_x(0,t) &= u_x(L,t) = 0, \\ u(x,0) &= f(x). \end{aligned}$$

First let $w = u - T$ to obtain the new problem

$$\begin{aligned} w_t &= kw_{xx} - Aw, \\ w_x(0,t) &= w_x(L,t) = 0, \\ w(x,0) &= f(x) - T. \end{aligned}$$

Now let $U(x,t) = e^{-At}w(x,t)$. The problem for $U(x,t)$ is

$$\begin{aligned} U_t &= kU_{xx}, \\ U_x(0,t) &= U_x(L,t) = 0, \\ U(x,0) &= w(x,0) = f(x) - T. \end{aligned}$$

The solution of this problem is

$$U(x,t) = \frac{1}{2}a_0 + \sum_{n=1}^{\infty} a_n \sin(n\pi x/L) e^{-kn^2\pi^2 t/L^2},$$

where

$$a_n = \frac{2}{L}\int_0^L (f(\xi) - T)\sin(n\pi\xi/L)\,d\xi.$$

Then

$$u(x,t) = w(x,t) + T = e^{-At}U(x,t) + T.$$

22. Multiply the heat equation by u and integrate to get

$$\int_a^b u u_t\,dx = k\int_a^b u u_{xx}\,dx.$$

Integrate the right side of this equation by parts and rewrite the left side as the integral of a partial derivative to obtain

$$\int_a^b \frac{1}{2}\frac{\partial}{\partial t}(u^2)\,dx = k\left([uu_{xx}]_a^b - \int_a^b u_x u_t\,dx\right).$$

This is equivalent to the equation to be derived.

2.2 A Nonhomogeneous Problem

2. First compute

$$\begin{aligned} B_n(t) &= \frac{2}{L}\int_0^L \xi\sin(t)\sin(n\pi\xi/L)\,d\xi \\ &= \frac{2L(-1)^{n+1}}{n\pi}\sin(t). \end{aligned}$$

Next, we need

$$\begin{aligned} &\frac{2L(-1)^{n+1}}{n\pi}\int_0^t e^{-kn^2\pi^2(t-\tau)/L^2}\sin(\tau)\,d\tau \\ &= \frac{2L^3(-1)^n}{n\pi(k^2n^4\pi^4 + L^4)}\left(L^2\cos(t) - L^2 e^{-kn^2\pi^2 t/L^2} - kn^2\pi^2\sin(t)\right). \end{aligned}$$

Denote this quantity $P_n(t)$.

Next, compute

$$b_n = \frac{2}{L}\int_0^L f(\xi)\sin(n\pi\xi/L)\,d\xi = \frac{2}{n\pi}(1-(-1)^n).$$

The solution is

$$\begin{aligned} u(x,t) &= \sum_{n=1}^{\infty} P_n(t)\sin(n\pi x/L) \\ &\quad + \sum_{n=1}^{\infty} \frac{2}{n\pi}(1-(-1)^n)\sin(n\pi x/L)e^{-kn^2\pi^2 t/L^2}. \end{aligned}$$

4. First,

$$B_n(t) = \int_0^{L/2} K\sin(n\pi\xi/L)\,d\xi = \frac{2K}{n\pi}\left(1-\cos(n\pi/2)\right).$$

Next,

$$\begin{aligned} P_n(t) &= \frac{2K}{n\pi}\left(1-\cos(n\pi/2)\right)\int_0^t e^{-kn^2\pi^2(t-\tau)/L^2}\,d\tau \\ &= \frac{2KL^2}{kn^3\pi^3}(1-\cos(n\pi/2))\left(1-e^{-kn^2\pi^2 t/L^2}\right). \end{aligned}$$

Next,

$$b_n = \frac{2}{L}\int_0^L \sin(\pi\xi/L)\sin(n\pi\xi/L)\,d\xi = \begin{cases} 0 & \text{if } n \neq 1, \\ 1 & \text{if } n = 1. \end{cases}$$

6. Attempt a solution

$$u(x,t) = \frac{1}{2}T_0(t) + \sum_{n=1}^{\infty} T_n(t)\cos(n\pi x/L).$$

Here

$$T_n(t) = \frac{2}{L}\int_0^L u(\xi,t)\sin(n\pi\xi/L)\,d\xi$$

for $n = 0, 1, 2, \ldots$. Expand,

$$F(x,t) = \frac{1}{2}A_0(t) + \sum_{n=1}^{\infty} A_n(t)\cos(n\pi x/L).$$

For any $t \geq 0$, this is the Fourier cosine expansion of $F(x,t)$ on $[0,L]$, thinking of $F(x,t)$ as a function of x. Therefore the coefficients are

$$A_n(t) = \frac{2}{L}\int_0^L F(\xi,t)\cos(n\pi\xi/L)\,d\xi.$$

Now differentiate $T_n(t)$ and use the heat equation to obtain

$$\begin{aligned}
T_n'(t) &= \frac{2}{L}\int_0^L u_t(\xi,t)\cos(n\pi\xi/L)\,d\xi \\
&= \frac{2}{L}\int_0^L (ku_{xx}(\xi,t)+F(\xi,t))\cos(n\pi\xi/L)\,d\xi \\
&= \frac{2k}{L}\int_0^L u_{xx}(\xi,t)\cos(n\pi\xi/L)\,d\xi \\
&\quad + \frac{2}{L}\int_0^L F(\xi,t)\cos(n\pi\xi/L)\,d\xi \\
&= \frac{2k}{L}\int_0^L u_{xx}(\xi,t)\cos(n\pi\xi/L) + A_n(t).
\end{aligned}$$

Integrate the last integral by parts and use the boundary conditions to obtain

$$T_n'(t) + \frac{kn^2\pi^2}{L^2}T_n(t) = A_n(t).$$

Now

$$\begin{aligned}
T_n(0) &= \frac{2}{L}\int_0^L u(\xi,t)\cos(n\pi\xi/L)\,d\xi \\
&= \frac{2}{L}\int_0^L f(\xi)\cos(n\pi\xi/L)\,d\xi = a_n,
\end{aligned}$$

the nth Fourier cosine coefficient of $f(x)$ on $[0,L]$. Thus $T_n(t)$ is determined as the solution of the problem

$$T_n'(t) + \frac{kn^2\pi^2}{L^2}T_n(t) = A_n(t); A_n(0) = a_n.$$

This has the unique solution

$$T_n(t) = \int_0^t e^{-kn^2\pi^2(t-\tau)/L^2}A_n(\tau)\,d\tau + a_n e^{-kn^2\pi^2 t/L^2},$$

where the a_n's are the Fourier cosine coefficients of $f(x)$ on $[0,L]$.

This results in the solution

$$\begin{aligned}
u(x,t) =& \frac{1}{2}T_0(t) + \sum_{n=1}^{\infty}\left(\int_0^t e^{-kn^2\pi^2(t-\tau)/L^2}A_n(\tau)\,d\tau\right)\cos(n\pi x/L) \\
&+ \frac{1}{2}a_0 + \sum_{n=1}^{\infty} a_n e^{-kn^2\pi^2 t/L^2}\cos(n\pi x/L).
\end{aligned}$$

7. Compute

$$A_0(t) = \frac{2}{L}\int_0^L \xi t\,d\xi = Lt$$

and for $n = 1, 2, \ldots,$

$$A_n = \frac{2}{L}\int_0^L \xi t \cos(n\pi\xi/L)\, d\xi = \frac{2L}{n^2\pi^2}(-1 + (-1)^n)t$$

Next, the Fourier cosine coefficients of $f(x) = 1$ on $[0, L]$ are

$$a_0 = \frac{2}{L}\int_0^L d\xi = 2,$$

$$a_n = \frac{2}{L}\int_0^L \cos(n\pi\xi/L)\, d\xi = 0 \text{ for } n = 1, 2, \ldots.$$

Let

$$P_0(t) = \int_0^t A_n(\tau)\, d\tau = \frac{L}{2}t^2$$

and for $n = 1, 2, \ldots,$

$$\begin{aligned} P_n(t) &= \frac{2L(-1 + (-1)^n)}{\pi^2 n^2}\int_0^t \tau e^{-kn^2\pi^2(t-\tau)/L}\, d\tau \\ &= -\frac{2L^2}{k^2 n^6 \pi^6}(1 - (-1)^n)\left[Le^{-kn^2\pi^2 t/L} - L + kn^2\pi^2 t\right]. \end{aligned}$$

Finally, the solution is

$$u(x,t) = \frac{L}{2}t^2 + \sum_{n=1}^{\infty} P_n(t)\cos(n\pi x/L) + 1.$$

8. Attempt a solution of the form

$$u(x,t) = \sum_{n=1}^{\infty} T_n(t)\sin((2n-1)\pi x/2L),$$

in which

$$T_n(t) = \frac{2}{L}\int_0^L u(\xi, t)\sin((2n-1)\pi\xi/2L)\, d\xi.$$

Carry out an analysis like that done in this section (substitute for $u_t(\xi, t)$ and integrate by parts, using the boundary conditions) to derive the expression

$$\begin{aligned} T_n'(t) &= \frac{2}{L}\int_0^L u_t(\xi,t)\sin((2n-1)\pi\xi/2L)\, d\xi \\ &= \frac{2k}{L}\beta(t)(-1)^{n+1} + \frac{2k}{L}\frac{(2n-1)\pi}{2L}\alpha(t) \\ &\quad - \frac{2k}{L}\left(\frac{(2n-1)\pi}{2L}\right)^2 \int_0^L u(\xi,t)\sin((2n-1)\pi\xi/2L)\, d\xi. \end{aligned}$$

Thus show that

$$T_n'(t) + k\lambda_n T_n(t) = b_n(t),$$

where

$$T_n(0) = 0 \text{ and } b_n(t) = \frac{2k}{L}\left(\sqrt{\lambda_n}\alpha(t) + (-1)^{n+1}\beta(t)\right).$$

9. With $\alpha(t) = 1$ and $\beta(t) = t$, attempt a solution

$$u(x,t) = \sum_{n=1}^{\infty} T_n(t)\sin((2n-1)\pi x/2L)$$

and let

$$\lambda_n = \left(\frac{(2n-1)\pi}{2L}\right)^2.$$

Solve

$$T_n'(t) + k\lambda_n T_n(t) = b_n(t), T_n(0) = 0,$$

where

$$b_n(t) = \frac{2k}{L}\sqrt{\lambda_n} + \frac{2k}{L}(-1)^{n+1}t.$$

Obtain

$$\begin{aligned} T_n(t) =& \frac{2}{L\sqrt{\lambda_n}}\left(1 - e^{-k\lambda_n t}\right) \\ &+ \frac{2(-1)^{n+1}}{kL\lambda_n^2}\left(k\lambda_n t - 1 + e^{-k\lambda_n t}\right). \end{aligned}$$

10. Attempt a solution of the form

$$u(x,t) = \frac{1}{2}T_0(t) + \sum_{n=1}^{\infty} T_n(t)\cos(n\pi x/L).$$

Chapter 3

Solutions of the Wave Equation

3.1 Solutions on Bounded Intervals

2.

$$u(x,t) = \sum_{n=1}^{\infty} \frac{(-1)^{n+1}}{n^2\pi^2} \sin(n\pi x)\sin(2n\pi t).$$

4. The solution reduces to a single term

$$u(x,t) = \sin(x)\cos(4t).$$

6.

$$u(x,t) = \sum_{n=1}^{\infty} \frac{-16}{\pi(2n-1)((2n-1)^2-4)} \sin((2n-1)x/2)\cos((4n-2)t).$$

8.

$$u(x,t) = \sum_{n=1}^{\infty} (a_n \cos(5n\pi t/2) + b_n \sin(5n\pi t/2))\sin(n\pi x/2),$$

where

$$a_n = -32\left(\frac{2(-1)^n+1}{n^3\pi^3}\right)$$

and

$$b_n = \frac{-16}{5n^4\pi^4}(n^2\pi^2(-1)^n - 2(-1)^n + 1).$$

Solutions Manual to Accompany Beginning Partial Differential Equations, Third Edition. Peter V. O'Neil.

10.

$$u(x,t) = \sum_{n=1}^{\infty} (a_n \cos(3n\pi t) + b_n \sin(3n\pi t)) \sin(n\pi x),$$

where

$$a_n = \frac{12(-1)^{n+1}}{(n\pi)^3} \text{ and } b_n = -\frac{2}{3}\left(\frac{\cos(1)(-1)^n - 1}{n^2\pi^2 - 1}\right).$$

14. Let $u(x,t) = X(x)T(t)$ in the telegraph equation to get

$$X'' + \lambda X = 0,\ T'' + AT' + (B + \lambda c^2)T = 0.$$

Because $X(0) = X(L) = 0$, the eigenvalues and eigenfunctions are

$$\lambda_n = \frac{n^2\pi^2}{L^2},\ X_n(x) = \sin\left(\frac{n\pi x}{L}\right).$$

For $T(t)$ we must solve

$$T'' + AT' + \left(B + \frac{n^2\pi^2c^2}{L^2}\right)T = 0;\ T'(0) = 0.$$

To obtain solutions $e^{\alpha t}$, substitute this into the differential equation and solve for α. To retain the dependence on n, denote the solutions for α as

$$\alpha_n = -\frac{A}{2} \pm \frac{1}{2}\sqrt{A^2 - 4\left(B + \frac{n^2\pi^2c^2}{L^2}\right)}.$$

By assumption the quantity under the radical is negative, so

$$\alpha_n = -\frac{A}{2} \pm \beta_n i,$$

where

$$\beta_n = \frac{1}{2}\sqrt{4\left(B + \frac{n^2\pi^2c^2}{L^2}\right) - A^2}.$$

Therefore, for $n = 1, 2, \ldots$, $T_n(t)$ has the form

$$T_n(t) = a_n e^{-At/2}\cos(\beta_n t) + b_n \sin(\beta_n t).$$

Now attempt a solution

$$u(x,t) = e^{-At/2}\sum_{n=1}^{\infty}(a_n \cos(\beta_n t) + b_n \sin(\beta_n t))\sin(n\pi x/L).$$

Now

$$u(x,0) = \sum_{n=1}^{\infty} a_n \sin(n\pi x/L) = \varphi(x),$$

so choose

$$a_n = \frac{2}{L}\int_0^L \varphi(\xi)\sin(n\pi\xi/L)\,d\xi.$$

Next,

$$u_t(x,0) = 0 = -\frac{A}{2}\sum_{n=1}^{\infty} a_n\beta_n \sin(n\pi x/L)$$
$$+\sum_{n=1}^{\infty} b_n\beta_n \sin(n\pi x/L).$$

Then

$$\sum_{n=1}^{\infty}\left(b_n\beta_n - \frac{A}{2}a_n\right)\sin(n\pi x/L) = 0$$

for $0 < x < L$. Then

$$\beta_n b_n - \frac{A}{2}a_n = 0$$

so

$$b_n = \frac{A}{L\beta_n}\int_0^L \varphi(\xi)\sin(n\pi\xi/L)\,d\xi.$$

16. Let $\theta(x,t) = X(x)T(t)$ and use the boundary conditions to obtain

$$X'' + \lambda X = 0,$$
$$X'(0) - \alpha X(0) = 0,$$
$$X'(L) + \alpha X(L) = 0,$$
$$T'' + \lambda\alpha^2 T = 0.$$

Consideration of cases on λ shows that 0 is not an eigenvalue, and there is no negative eigenvalue. Set $\lambda = k^2$ with $k > 0$ to obtain solutions for X of the form

$$X(x) = c\cos(kx) + d\sin(kx).$$

From the boundary conditions we obtain

$$kd - \alpha c = 0$$

and

$$-kc\sin(kL) + kd\cos(kL) + \alpha(c\cos(kL) + d\sin(kL)) = 0.$$

From these we obtain

$$\tan(kL) = \frac{2\alpha k}{j^2 - \alpha^2}.$$

If we think of the left and right sides of this equation as functions of k, the straight line graph (right side) intersects the graph of the tangent function (right side) infinitely many times with $k > 0$. The first coordinate of each such point is an eigenvalue of this problem. If k_n is the nth such first coordinate (counting from left to right), then the eigenvalues are $\lambda_n = k_n^2$. Although we cannot solve for k_n in an exact algebraic expression, we can approximate these numbers to any degree of accuracy we need. The problem for T is now

$$T'' + \alpha^2 k_n^2 T = 0.$$

The condition $\theta_t(x, 0) = 0$ implies that $T'(0) = 0$. Therefore $T_n(t)$ is a constant multiple of $\cos(\alpha k_n t)$ and we have functions

$$\begin{aligned}\theta_n(x,t) &= (c_n \cos(k_n x) + d_n \sin(k_n x)) \cos(\alpha k_n t)\\ &= d_n \left(\frac{k_n}{\alpha} \cos(k_n x) + \sin(k_n x)\right) \cos(\alpha k_n t),\end{aligned}$$

which satisfy the partial differential equation, both boundary conditions, and the zero initial velocity condition. To satisfy $u(x, 0) = \varphi(x)$, attempt a superposition

$$\theta(x,t) = \sum_{n=1}^{\infty} \theta_n(x,t).$$

The coefficients d_n must be chosen so that

$$\theta(x,0) = \sum_{n=1}^{\infty} d_n \left(\frac{k_n}{\alpha} \cos(\alpha x) + \sin(\alpha x)\right) = \varphi(x).$$

This reminds one of a Fourier series, but here the functions we are expanding $\varphi(x)$ in terms of are

$$f_n(x) = \frac{k_n}{\alpha} \cos(k_n x) + \sin(k_n x).$$

However, using the transcendental equation defining the numbers k_n, it is easy to show that

$$\int_0^L f_n(x) f_m(x)\, dx = 0 \text{ if } n \neq m.$$

Multiply the proposed expansion by $f_m(x)$ to obtain

$$\varphi(x) f_m(x) = \sum_{n=1}^{\infty} d_n f_n(x) f_m(x).$$

Upon integrating term by term, all terms on the right vanish except possibly the $n = m$ term, yielding

$$d_m = \frac{\int_0^L \varphi(x) f_m\, dx}{\int_0^L f_m^2(x)\, dx}.$$

18. Let $u(x,t) = v(x,t) + f(x)$ and substitute into the wave equation to obtain

$$v_{tt} = 9(v_{xx} + f''(x)) + Ax^2.$$

Thus choose $f(x)$ so that

$$9f''(x) + Ax^2 = 0.$$

Integrate twice to get

$$f(x) = -\frac{A}{108}x^4 + Cx + D.$$

Now, the condition

$$u(0,t) = v(0,t) + f(0) = v(0,t) + D = 0$$

becomes just $v(0,t) = 0$ if we choose $D = 0$. Further,

$$u(1,t) = v(1,t) + f(1) = 0$$

becomes $v(1,t) = 0$ if we choose C so that $f(1) = 0$. Thus choose

$$C = \frac{A}{108}$$

so

$$f(x) = \frac{A}{108}x(1 - x^3).$$

Next, we will have

$$u(x,0) = v(x,0) + f(x) = 0$$

if we require that $v(x,0) = -f(x)$. Finally,

$$u_t(x,0) = v_t(x,0) = 0.$$

This familiar problem for $v(x,t)$ has the solution

$$v(x,t) = \sum_{n=1}^{\infty} c_n \cos(3nt)\sin(n\pi x),$$

where

$$\begin{aligned} c_n &= 2\int_0^1 -f(\xi)\sin(n\pi\xi)\,d\xi \\ &= -2\int_0^1 \frac{A}{108}\xi(1-\xi^3)\sin(n\pi\xi)\,d\xi \\ &= \frac{A}{9n^5\pi^5}\left(4(1-(-1)^n) + 2n^2\pi^2(-1)^n\right). \end{aligned}$$

Then $u(x,t) = v(x,t) + f(x)$.

20. Suppose α is a positive number that is not an integer. Let $u(x,t) = v(x,t) + f(x)$ to obtain the solution

$$u(x,t) = \sum_{n=1}^{\infty} c_n \cos(2nt)\sin(nx) + \frac{1}{4\alpha^2}\cos(\alpha x) + \frac{1}{4\alpha^2}(1-\cos(\alpha\pi))x - \frac{1}{4\alpha^2},$$

where

$$c_n = -\frac{n}{2\pi}\frac{\cos(\alpha\pi)(-1)^n - 1}{\alpha^2(n^2-\alpha^2)} + \frac{n(-1)^n}{2}\frac{\cos(\alpha\pi)-1}{n^2\pi^2} + \frac{1}{2\pi}\frac{(-1)^n - 1}{n\alpha^2}.$$

A different solution must be derived if α is a positive integer.

22. The solution is

$$u(x,t) = \sum_{n=1}^{\infty} a_n \cos(4n\pi t/3)\sin(n\pi x/3) + \frac{1}{16}(e^{-x} - 1) + \frac{1}{48}(1-e^{-3})x,$$

where

$$a_n = \frac{1}{8n\pi}(1 - e^{-3}(-1)^n) + \frac{n\pi}{8(9+n^2\pi^2)}((-1)^n - 1).$$

24. The solution is

$$u(x,t) = \sum_{n=1}^{\infty}\left(\frac{2(-1)^{n+1}}{n\pi}\cos(2n\pi t/9) - \frac{81(-1)^n}{n^2\pi^2}\sin(2n\pi t/9)\right)\sin(n\pi x/9) + \frac{1}{9}x.$$

26. Multiply the partial differential equation by u_t to get

$$u_t u_{tt} = c^2 u_t u_{xx} + u_t g(x,t).$$

Then

$$\int_a^b u_t u_{tt}\,dx = \int c^2 u_t u_{xx}\,dx + \int_a^b u_t g(x,t)\,dx.$$

But

$$u_t u_{tt} = \frac{\partial}{\partial t}\left(\frac{1}{2}u_t^2\right).$$

Therefore

$$\frac{d}{dt}\int_a^b \frac{1}{2}u_t^2\,dx = \int_a^b c^2 u_t u_{xx}\,dx + \int_a^b u_t g(x,t)\,dx.$$

Integrate by parts to get

$$\begin{aligned}\int_a^b c^2 u_t u_{xx}\,dx &= c^2\left[u_t u_x\right]_a^b - c^2\int_a^b u_x u_{xt}\,dx\\ &= c^2\left[u_t u_x\right]_a^b - c^2\int_a^b \frac{\partial}{\partial t}\left(\frac{1}{2}u_x^2\right)\,dx.\end{aligned}$$

Then

$$\frac{d}{dt}\int_a^b \frac{1}{2}u_t^2\,dx = c^2\left[u_t u_x\right]_a^b - c^2\int_a^b \frac{\partial}{\partial t}\left(\frac{1}{2}u_x^2\right)\,dx + \int_a^b u_t g(x,t)\,dx.$$

Rearrangement of this equation yields the conclusion to be proved.

27. Let $u(x,t)$ and $v(x,t)$ be solutions and let $w(x,t) = u(x,t) - v(x,t)$. Then w is a solution of the problem

$$\begin{aligned}w_{tt} &= c^2 w_{xx},\\ w_x(0,t) &= w_x(L,t) = 0,\\ w(x,0) &= w_t(x,0) = 0.\end{aligned}$$

Define

$$E(t) = \frac{1}{2}\int_0^L (w_t^2 + c^2 w_x^2)\,dx.$$

Then

$$E'(t) = \int_0^L (w_t w_{tt} + c^2 w_x w_{xt})\,dx.$$

Integrate the second term by parts and use the boundary conditions to conclude that

$$\begin{aligned}\int_0^L w_x w_{xt}\,dx &= w_x w_t\big]_0^L - \int_0^L w_t w_{xx}\,dx\\ &= -\int_0^L w_t w_{xx}\,dx.\end{aligned}$$

Now use the partial differential equation to write

$$\begin{aligned}E'(t) &= \int_0^L (w_t w_{tt} - c^2 w_t w_{xx})\,dx\\ &= \int_0^L \int_0^L (w_t w_{tt} - w_t w_{tt})\,dx = 0\end{aligned}$$

for $t > 0$ and $0 < x < L$. Because $E(t)$ is continuous, $E(t)$ is constant on any interval $[0,T]$. But $E(0) = 0$ so $E(t)$ is identically zero, and therefore w_x and w_t must be zero. This means that $w(x,t)$ must be constant. But $w(x,0) = 0$, so $w(x,t) = 0$ and $u(x,t) = v(x,t)$.

28.

$$u(x,t) = e^{-t/2}\left(\cos(\sqrt{35}t) + \frac{1}{2\sqrt{35}}\sin(\sqrt{35}t)\right)\sin(x).$$

30.

$$\begin{aligned} u(x,t) =& \frac{\pi}{2}e^{-t/2}\left(\cos(4\sqrt{63}t) + \frac{1}{2\sqrt{63t}}\sin(4\sqrt{63}t)\right)\sin(x) \\ &+ e^{-t/2}\sum_{n=2}^{\infty} b_n \left(\cos(4\sqrt{63}t) + \frac{1}{2\sqrt{63}}\sin(4\sqrt{63}t)\right)\sin(nx), \end{aligned}$$

where

$$b_n = \frac{-4n(1+(-1)^n)}{\pi(n^2-1)^2} \text{ for } n = 2, 3, \ldots.$$

32.

$$u(x,t) = e^{-t/8}\sum_{n=1}^{\infty} b_n \left(\cos(\alpha_n t) + \frac{2}{\alpha_n}\sin(\alpha_n t)\right)\sin(nx),$$

where

$$\alpha_n = \frac{1}{2}\sqrt{64n^2 - 1}$$

and

$$b_n = \frac{4}{n^3}(2 + (-1)^n).$$

3.2 The Cauchy Problem

3.2.1 d'Alembert's Solution

2.

$$\begin{aligned} u(x,t) &= \frac{1}{2}\left((x+4t)^2 + (x-4t)^2\right) + \frac{1}{8}\int_{x-4t}^{x+4t}\sin(2s)\,ds \\ &= x^2 + 16t^2 + \frac{1}{8}\sin(2x)\sin(8t). \end{aligned}$$

4.

$$u(x,t) = \frac{1}{2}(\cosh(x+2t) + \cosh(x-2t)) + 2xt.$$

6.

$$u(x,t) = 2 + x + \frac{1}{2}\left(e^{x+t} - e^{x-t}\right).$$

8.

$$\begin{aligned} u(x,t) &= \frac{1}{2}(\sin(3(x+t)) + \sin(3(x-t))) + \frac{1}{2}\int_{x-t}^{x+t}\cos(s)\,ds \\ &= \frac{1}{2}(\sin(x+t) - \sin(x-t)). \end{aligned}$$

11. The solution with $\varphi(x) = \sin(x)$ is

$$u(x,t) = \frac{1}{2}(\sin(x+t) + \sin(x-t))$$

while the problem with $\varphi(x) = \sin(x) + \epsilon$ has the solution $u(x,t) + \epsilon$.

12. The solution of the first problem is

$$u_1(x,t) = \frac{1}{2}(\cos(x+t) + \cos(x-t)) + xt.$$

The solution of the second problem is

$$u_2(x,t) = \frac{1}{2}(\cos(x+t) + \cos(x-t)) + xt + \epsilon t + \epsilon.$$

Then

$$|u_2(x,t) - u_1(x,t)| = \epsilon(1+t).$$

On any interval $0 \le x \le T$, this difference has magnitude not exceeding $\epsilon(1+T)$.

13. Let $v(x,t) = \int_0^t w(x,t,T)\,dT$ and show that u is a solution of the Cauchy problem with the given initial conditions. Compute

$$v_t = w(x,t,t) + \int_0^t w_t(x,t,T)\,dT = \int_0^t w_t(x,t,T)\,dT$$

because $w(x,t,t) = 0$. Similarly,

$$\begin{aligned} v_{tt} &= w_t(x,t,t) + \int_0^t w_{tt}(x,t,T)\,dT \\ &= f(x,t) + \int_0^t w_{tt}(x,t,T)\,dT, \\ v_x &= \int_0^t w_x(x,t,T)\,dT, \\ v_{xx} &= \int_0^t w_{xx}(x,t,T)\,dT. \end{aligned}$$

Then

$$v_{tt} - v_{xx} = f(x,t) + \int_0^t (w_{tt}(x,t,T) - w_{xx}(x,t,T))\,dT = f(x,t)$$

because $w_{tt} = w_{xx}$. This shows that $v(x,t)$ satisfies the partial differential equation. Finally,

$$v(x,0) = 0 = v_t(x,0).$$

Therefore $v(x,t)$ is a solution of the Cauchy problem. Since this solution is unique, then

$$v(x,t) = u(x,t) = \int_0^t w(x,t,T)\,dT.$$

16. With zero initial velocity, the solution is

$$u(x,t) = \frac{1}{2}(\varphi(x-t) + \varphi(x+t)),$$

the sum of a forward and backward wave, respectively. Figures 3.1–3.5 show the wave at times $t = 0$, $t = 0.3$, $t = 0.6$, $t = 1$, and 1.3, respectively. At $t = 1.3$ the forward and backward waves have separated.

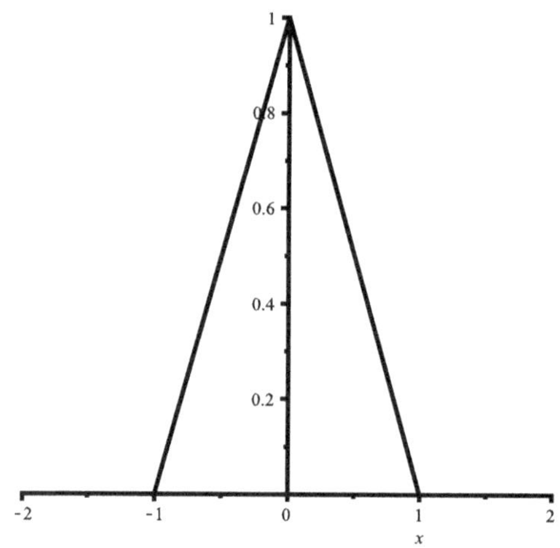

Figure 3.1: Problem 16, wave at time $t = 0$.

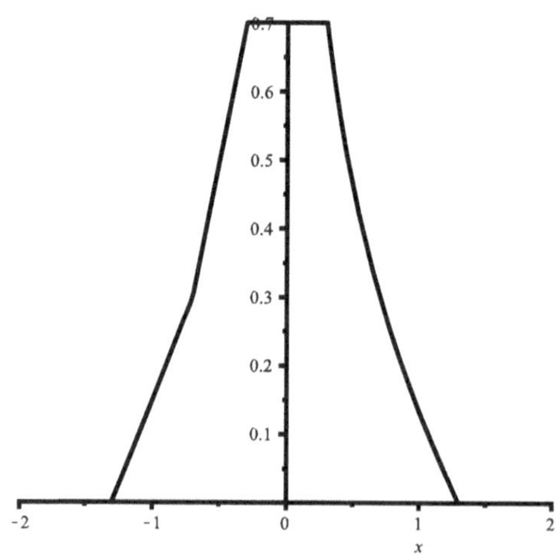

Figure 3.2: Problem 16, wave at time $t = 0.3$.

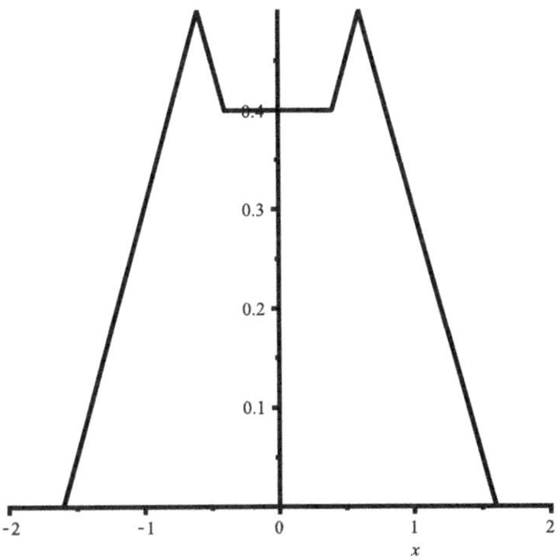

Figure 3.3: Problem 16, wave at time $t = 0.6$.

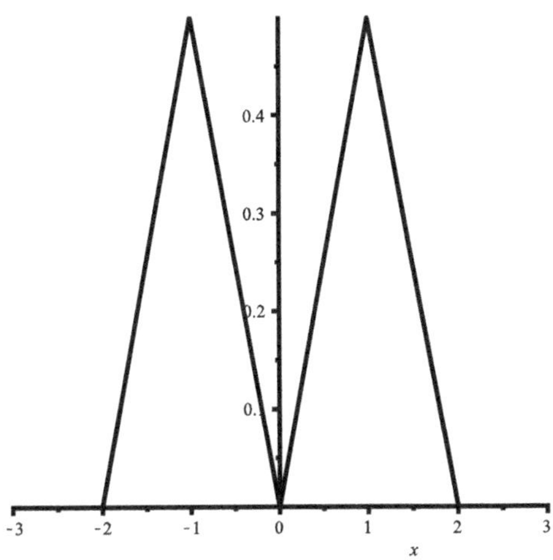

Figure 3.4: Problem 16, wave at time $t = 1$.

18. Figures 3.6–3.9 show the wave at times $t = 0$, $t = 0.3$, $t = 0.7$, and $t = 1.3$, respectively.

20. Figures 3.10–3.13 show the wave at times $t = 0$, $t = 0.3$, $t = 0.6$, and $t = 1.3$.

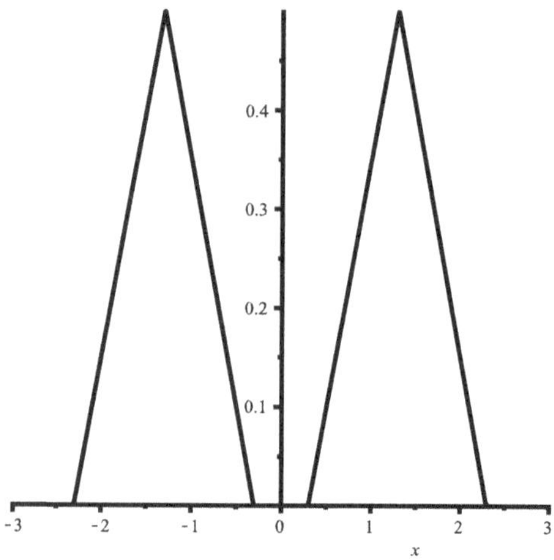

Figure 3.5: Problem 16, wave at time $t = 1.3$.

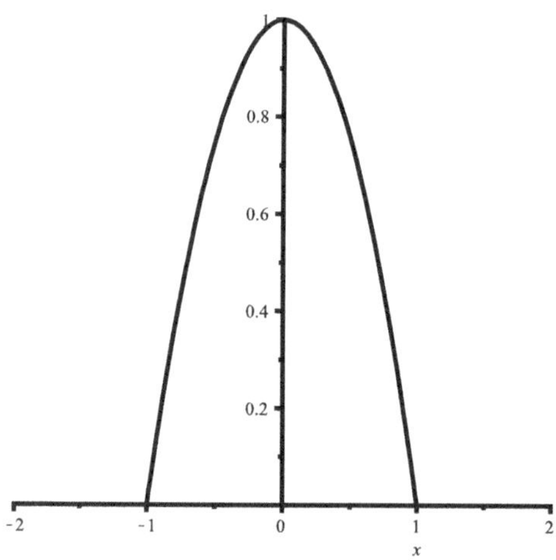

Figure 3.6: Problem 18, wave at time $t = 0$.

3.2.2 The Cauchy Problem on a Half Line

2. First make odd extensions of $\varphi(x)$ and $\psi(x)$:

$$\Phi(x) = \begin{cases} x^2 & \text{for } x \geq 0, \\ -x^2 & \text{for } x < 0 \end{cases}$$

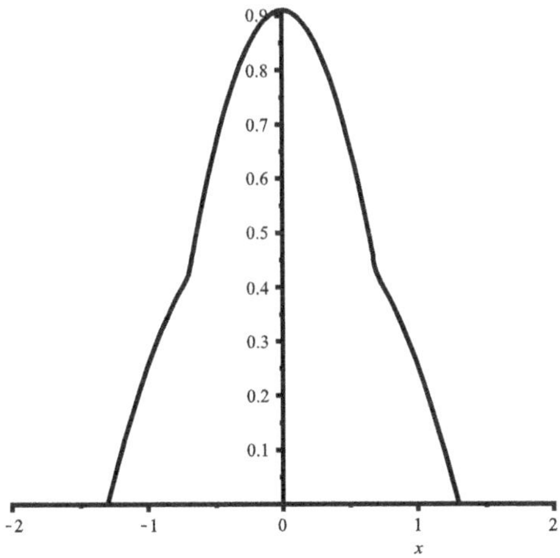

Figure 3.7: Problem 18, wave at time $t = 0.3$.

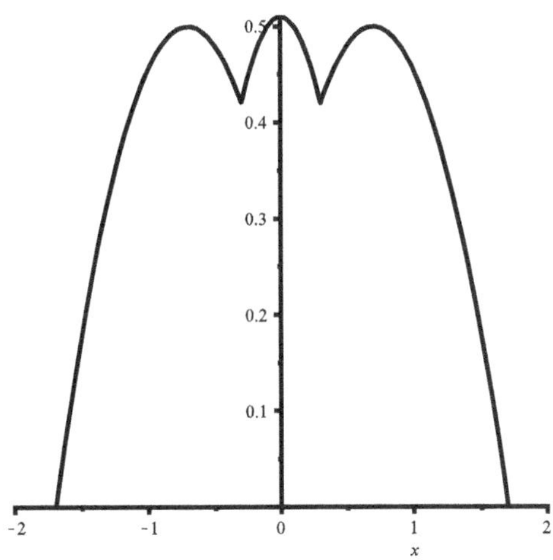

Figure 3.8: Problem 18, wave at time $t = 0.7$.

and $\Psi(x) = x$ for all real x. The solution is

$$u(x,t) = \frac{1}{2}\left(\Phi(x+4t) + \Phi(x-4t)\right) + \frac{1}{8}\int_{x-4t}^{x+4t} \Psi(s)\,ds.$$

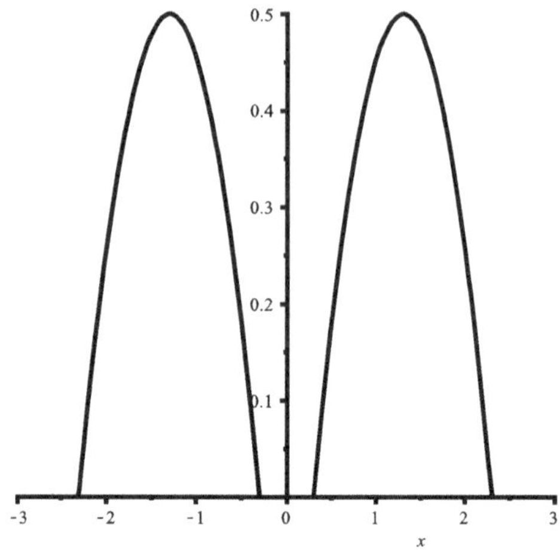

Figure 3.9: Problem 18, wave at time $t = 1.3$.

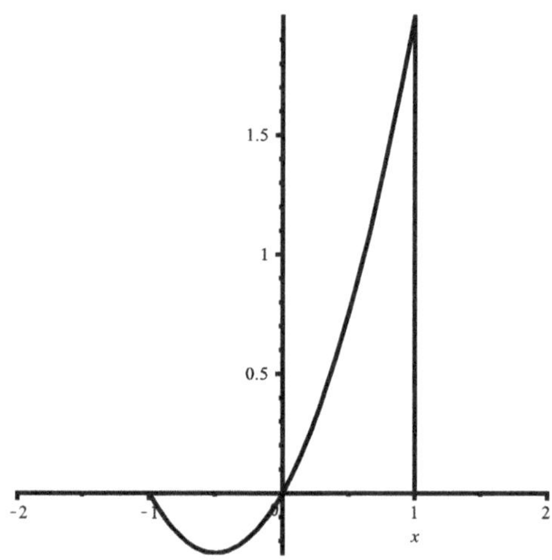

Figure 3.10: Problem 20, wave at time $t = 0$.

By taking cases we can write

$$u(x,t) = \frac{1}{2}((x+4t)^2 + (x-4t)^2) + \frac{1}{8}\int_{x-4t}^{x+4t} s\,ds$$

$$= x^2 + 16t^2 + xt \text{ for } x - 4t \geq 0$$

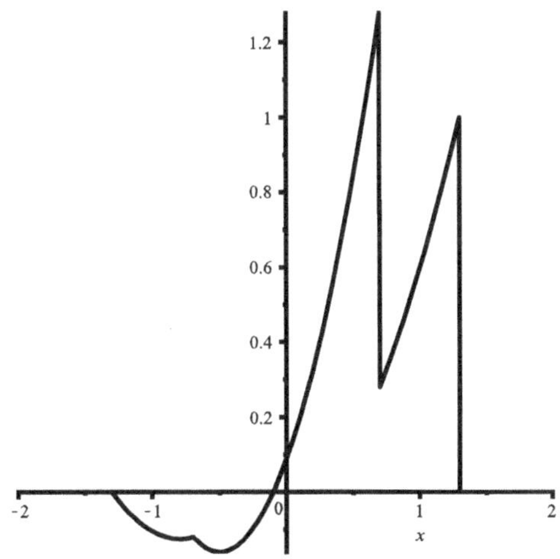

Figure 3.11: Problem 20, wave at time $t = 0.3$.

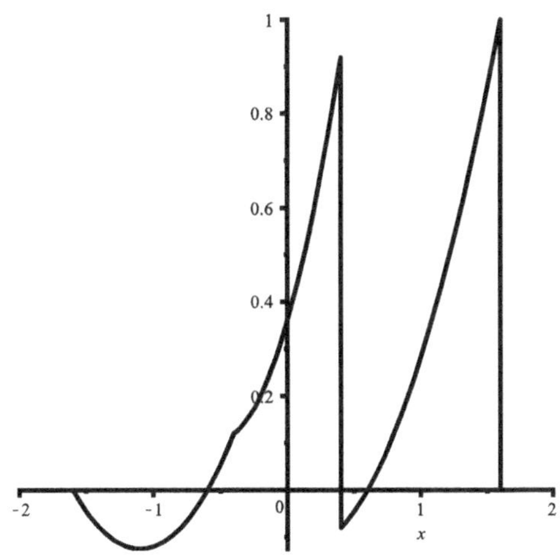

Figure 3.12: Problem 20, wave at time $t = 0.6$.

and

$$u(x,t) = \frac{1}{2}((x+4t)^2 - (x-4t)^2) + \frac{1}{8}\int_{x-4t}^{x+4t} s\,ds$$

$$= 9xt \text{ for } x - 4t < 0.$$

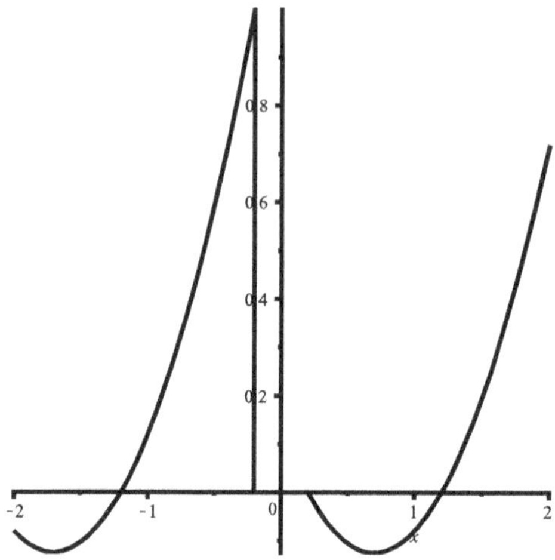

Figure 3.13: Problem 20, wave at time $t = 1.3$.

4. Make the odd extensions

$$\Phi(x) = \begin{cases} 1 - \cos(x) & \text{for } x \geq 0, \\ \cos(x) - 1 & \text{for } x < 0 \end{cases}$$

and

$$\Phi(x) = \begin{cases} e^{-x} & \text{for } x \geq 0, \\ -e^{x} & \text{for } x < 0. \end{cases}$$

Now obtain

$$u(x,t) = 1 - \cos(x)\cos(4t) + \frac{1}{4}e^{-x}\sinh(4t) \text{ if } x - 4t \geq 0$$

and

$$u(x,t) = \sin(x)\sin(4t) + \frac{1}{4}e^{-4t}\sinh(x) \text{ if } x - 4t < 0.$$

6. By making odd extensions of the initial position and velocity functions, obtain

$$u(x,t) = \frac{1}{2}(\sinh^2(x+7t) + \sinh^2(x-7t)) + xt \text{ for } x - t \geq 0$$

and

$$u(x,t) = \frac{1}{2}(\sinh^2(x+7t) - \sinh^2(x-7t)) + xt \text{ for } x - 7t < 0.$$

8. The solution is

$$u(x,t) = \begin{cases} x^2 + 9t^2 - x + xt - t & \text{for } x - 3t \geq 0, \\ 7xt - 4x/3 & \text{for } x - 3t < 0. \end{cases}$$

10.

$$u(x,t) = \begin{cases} x^3 + 108xt^2 + x + \frac{1}{12}\cos(2x)\sin(12t) & \text{for } x - 6t \geq 0, \\ x^3 + 108xt^2 + \frac{1}{12}\sin(2x)\cos(12t) & \text{for } x - 6t < 0. \end{cases}$$

3.2.3 Characteristic Triangles and Quadrilaterals

1. The solution is

$$u(x,t) = \frac{1}{2}(\varphi(x+ct) + \varphi(x-ct)) + \frac{1}{2c}\int_{x-ct}^{x+ct} \psi(s)\,ds.$$

Suppose $t > 0$ and $x > a + ct$. Then $x = a + ct + h$ for some positive number h. Then

$$\varphi(x+ct) = \varphi(a+2ct+h) = 0$$

because $a + 2ct + h > a$. Further,

$$\varphi(x-ct) = \varphi(a+h) = 0$$

because $a + h > a$. Finally,

$$\int_{x-ct}^{x+ct} \psi(s)\,ds = \int_{a+h}^{a+2ct+h} \psi(s)\,ds = 0$$

because $\psi(s) = 0$ on $[a+h, a+2ct+h]$. Therefore $u(x,t) = 0$ for $x > a+ct$. A similar argument shows that the solution vanishes if $x < -a - ct$.

3.2.4 A Cauchy Problem with a Forcing Term

2.

$$\begin{aligned} u(x,t) &= \frac{1}{2}(\sin(x+2t) + \sin(x-2t)) + \frac{1}{4}\int_{x-2t}^{x+2t} 2s\,ds \\ &\quad + \frac{1}{4}\int_0^t \int_{x-2t+2Y}^{x+2t-2Y} 2XY\,dX\,dY \\ &= \sin(x)\cos(2t) + 2xt + \frac{1}{3}xt^3. \end{aligned}$$

4.

$$\begin{aligned} u(x,t) &= \frac{1}{2}\left((x+4t)^2 + (x-4t)^2\right) + \frac{1}{8}\int_{x-4t}^{x+4t} se^{-s}\,ds \\ &\quad + \frac{1}{8}\int_0^t \int_{x-4t+4Y}^{x+4t-4Y} X\sin(Y)\,dX\,dY \\ &= x^2 + 16t^2 + \frac{1}{4}xe^{-x}\sinh(4t) - te^{-x}\cosh(4t) \\ &\quad + \frac{1}{4}e^{-x}\sinh(4t) - x\sin(t) + xt. \end{aligned}$$

6.

$$\begin{aligned}
u(x,t) &= \frac{1}{2}(1+x+7t+1+x-7t) + \frac{1}{14}\int_{x-7t}^{x+7t} \sin(s)\,ds \\
&+ \frac{1}{14}\int_0^t \int_{x-7t+7Y}^{x+7t-7Y} (X-\cos(Y))\,dX\,dY \\
&= 1+x+\frac{1}{14}(\cos(x-7t)-\cos(x+7t)) \\
&+ \frac{1}{2}xt^2 + \cos(t) + 1.
\end{aligned}$$

8.

$$\begin{aligned}
u(x,t) &= \frac{1}{2}\left((x+t)^2+(x-t)^2\right) + \frac{1}{2}\int_{x-t}^{x+t} \sin^2(s)\,ds \\
&+ \frac{1}{2}\int_0^t \int_{x-t+Y}^{x+t-Y} e^{-X}\cos(Y)\,dX\,dY \\
&= x^3+3xt^2-\cos^2(x)\cos(t)\sin(t)+\frac{1}{2}\sin(t)\cos(t) \\
&+ \frac{1}{2}t - \frac{1}{2}e^{-x}\cos(t) + \frac{1}{4}e^{-(x-t)} + \frac{1}{4}e^{-(x+t)}.
\end{aligned}$$

10.

$$\begin{aligned}
u(x,t) &= \frac{1}{2}(1-(x+6t)^2+1-(x-6t)^2) + \frac{1}{12}\int_{x-6t}^{x+6t} s\sin(s)\,ds \\
&+ \frac{1}{12}\int_0^t \int_{x-6t+6Y}^{x+6t-6Y} Y\sin(X)\,dX\,dY \\
&= 1-x^2-36t^2 \\
&+ \frac{1}{12}((x-6t)\cos(x-6t)-(x+6t)\cos(x+6t)) \\
&+ \frac{1}{12}(\sin(x+6t)-\sin(x-6t)) \\
&+ \frac{4}{27}\sin(x)\sin^3(t)\cos^3(t) \\
&- \frac{1}{36}\sin(x)\sin(t)\cos(t) + \frac{1}{36}t\sin(x).
\end{aligned}$$

3.2.5 String with Moving Ends

2. It is routine to verify that the compatibility conditions are satisfied. In region I, d'Alembert's formula applies and the solution is

$$\begin{aligned}
u(x,t) &= \frac{1}{2}(\varphi(x+2t)+\varphi(x-2t)) \\
&= \frac{1}{2}(\sin(\pi(x+2t))+\sin(\pi(x-2t))) \\
&= \sin(\pi x)\cos(2\pi t).
\end{aligned}$$

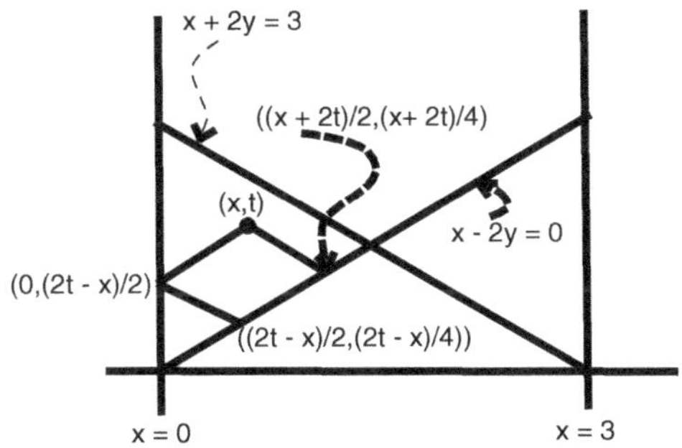

Figure 3.14: Region II for Problem 2.

For a point (x, t) in region II, Figure 3.14 shows a typical characteristic quadrilateral (to identify points and lines, but not drawn to scale). The solution at (x, t) is

$$\begin{aligned} u(x,t) &= u\left(0, \frac{2t-x}{2}\right) + u\left(\frac{x+2t}{2}, \frac{x+2t}{4}\right) - u\left(\frac{2t-x}{2}, \frac{2t-x}{4}\right) \\ &= \left(\frac{2t-x}{2}\right)^3 + \sin\left(\pi\left(\frac{x+2t}{2}\right)\right)\cos\left(\pi\left(\frac{x+2t}{2}\right)\right) \\ &\quad - \sin\left(\pi\left(\frac{2t-x}{2}\right)\right)\cos\left(\pi\left(\frac{2t-x}{2}\right)\right). \end{aligned}$$

Figure 3.15 shows a characteristic quadrilateral at (x, t) in region III. The solution at this point is

$$\begin{aligned} u(x,t) &= \sin\left(\pi\left(\frac{3+x-2t}{2}\right)\right)\cos\left(\pi\left(\frac{3-x+2t}{2}\right)\right) \\ &\quad + \left(\frac{-3+x+2t}{2}\right)^3 \\ &\quad - \sin\left(\pi\left(\frac{9-x-2t}{2}\right)\right)\cos\left(\pi\left(\frac{-3+x+2t}{2}\right)\right). \end{aligned}$$

4. Notice that

$$a''(0) = 12 \neq \varphi''(0),$$

so this problem has no solution.

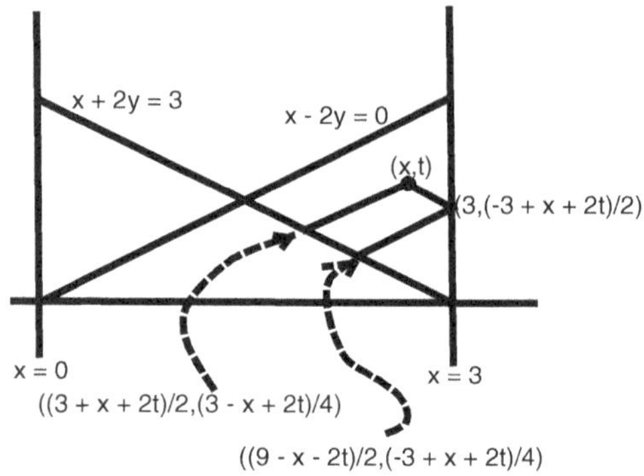

Figure 3.15: Region III for Problem 2.

6. The compatibility conditions are satisfied. At a point (x,t) in region I, the solution is

$$\begin{aligned}u(x,t) &= \frac{1}{2}\left((x+2t)\cos(\pi(x+2t)) + (x-2t)\cos(\pi(x-2t))\right) \\ &\quad + x - 2t + \frac{1}{4}\int_{x-2t}^{x+2t} x^2\,ds \\ &= \frac{1}{2}\left((x+2t)\cos(\pi(x+2t)) + (x-2t)\cos(\pi(x-2t))\right) + x + x^2 t + \frac{4}{3}t^3.\end{aligned}$$

At a point (x,t) in region II (Figure 3.16), the solution is

$$\begin{aligned}u(x,t) =& u\left(0,\frac{2t-x}{2}\right) + u\left(\frac{x+2t}{2},\frac{x+2t}{4}\right) \\ &- u\left(\frac{2t-x}{2},\frac{2t-x}{4}\right).\end{aligned}$$

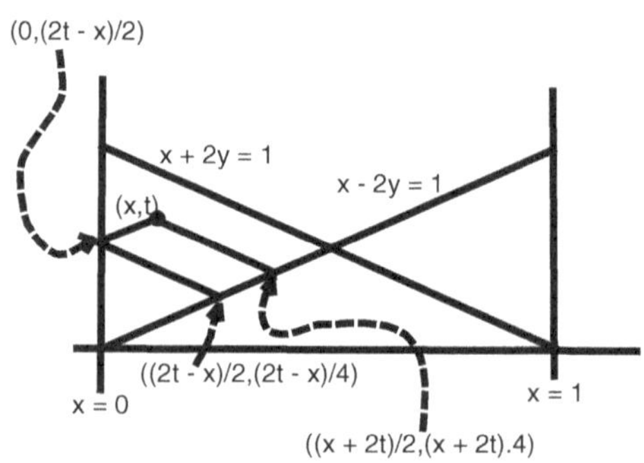

Figure 3.16: Region II for Problem 6.

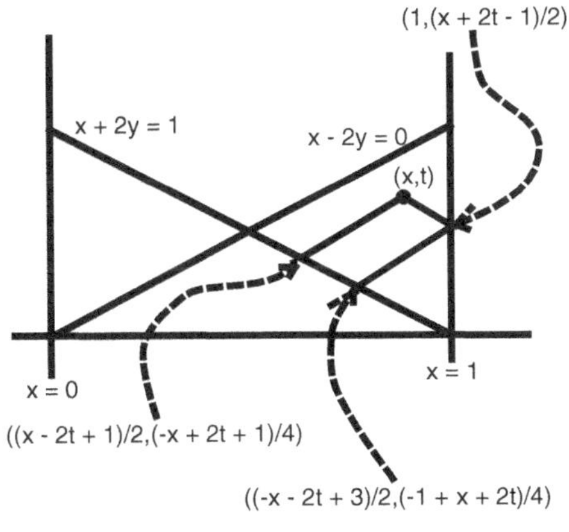

Figure 3.17: Region III for Problem 6.

After some manipulation, we obtain

$$\begin{aligned} u(x,t) =& x + \frac{1}{6}x^3 + 2xt^2 \\ & + \frac{1}{2}(x+2t)\cos(\pi(x+2t)) + \frac{1}{2}(x-2t)\cos(\pi(x-2t)). \end{aligned}$$

For (x,t) in region III (Figure 3.17),

$$\begin{aligned} u(x,t) &= u\left(\frac{x-2t+1}{2}, \frac{-x+2t+1}{4}\right) \\ &\quad + u\left(1, \frac{x+2t-1}{2}\right) - u\left(\frac{-x-2t+3}{2}, \frac{-1+x+2t}{4}\right) \\ &= -t - \frac{4}{3} - \frac{1}{6}xt^2 + \frac{1}{2}x^2 + 2xt + 2t^2 + \frac{1}{2}x \\ &\quad + 2\pi\left(\frac{x+2t-1}{2}\right)^2 + \frac{1}{2}(x-2t)\cos(\pi(x-2t)). \end{aligned}$$

For (x,t) in region IV (Figure 3.18), the solution is

$$\begin{aligned} u(x,t) &= u\left(\frac{x-2t+1}{2}, \frac{-x+2t+1}{4}\right) \\ &\quad - u\left(\frac{1}{2}, \frac{1}{4}\right) + u\left(\frac{x+2t}{2}, \frac{x+2t}{4}\right) \\ &= -t + \frac{1}{2}x - x^2t - \frac{4}{3}t^3 - \frac{4}{3} \\ &\quad + \frac{1}{2}x^2 + 2xt + 2t^2 + \frac{\pi}{2}(x+2t-1)^2 + \frac{1}{2}(x-2t)\cos(\pi(x-2t)). \end{aligned}$$

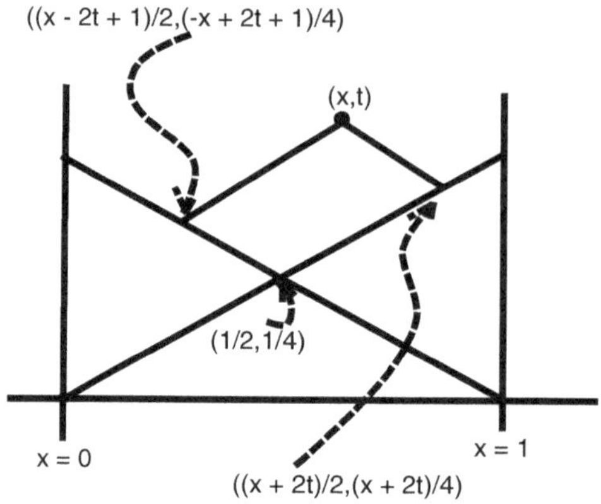

Figure 3.18: Region IV for Problem 6.

7. Immediately,

$$a(0) = u(0,0) = \varphi(0).$$

Next,

$$a'(t) = u_t(0,t),$$

so

$$a'(0) = u_t(0,0) = \psi(0).$$

Next,

$$a''(t) = c^2 u_{tt}(0,t) = c^2 u_{xx}(0,t),$$

so

$$a''(0) = c^2 u_{xx}(0,0) = c^2 \varphi''(0).$$

Similar calculations apply at the end $= L$.

3.3 The Wave Equation in Higher Dimensions

3.3.1 Vibrations in a Membrane with Fixed Frame

2. A separation of variables leads to the solution

$$u(x,y,t) =$$
$$\sum_{n=1}^{\infty}\sum_{m=1}^{\infty} \frac{128}{(nm)^3} \left(2(-1)^n+1\right)\left(2(-1)^m+1\right)\sin(nx)\sin(my/2)\cos(2\alpha_{nm}t),$$

where

$$\alpha_{nm} = \sqrt{n^2 + \frac{m^2}{4}}.$$

4. A separation of variables leads to the solution

$$u(x,y,t) = \frac{1}{4\sqrt{2}\pi}\sin(\pi x)\sin(\pi y)\sin(4\sqrt{2}\pi t).$$

6. The solution is

$$u(x,y,t) =$$
$$\sum_{n=1}^{\infty}\sum_{m=1}^{\infty}[a_{nm}\cos(\alpha_{nm}t) + b_{nm}\sin(\alpha_{nm}t)]\sin(n\pi x)\sin(m\pi y),$$

where

$$a_{nm} = \frac{64}{\pi^3}\frac{n(-1)^{n+1}(1-(-1)^n)}{m(4n^2-1)^2},$$
$$b_{nm} = \frac{4}{mn\pi\alpha_{nm}}[(-1)^{n+m}(1+\pi) - \pi(-1)^m - (-1)^n],$$

and

$$\alpha_{nm} = \sqrt{n^2\pi^2 + m^2}.$$

3.3.2 The Poisson Integral Solution

1. If $\varphi(x) = 0$ and $\psi(x) = k$, the solution of the Cauchy problem is

$$u(\mathbf{x},t) = \frac{1}{4\pi t}\iint_{S_t} k\,d\sigma_t = \frac{k}{4\pi t}(4\pi t^2) = kt,$$

and this is independent of $\mathbf{x}$.

3.3.3 Hadamard's Method of Descent

1. If ψ is a function of just one space variable x, then the solution of problem 3.25 is

$$u(x,t) = \frac{1}{4\pi t}\iint_{S_{(x,0,0;t)}}\psi(\xi)\,d\sigma_t = \frac{1}{2}\int_{x-t}^{x+t}\psi(\xi)\,d\xi.$$

This gives the solution of the Cauchy problem as

$$\begin{aligned} u(x,t) &= \frac{\partial}{\partial t}\left[\frac{1}{2}\int_{x-t}^{x+t}\varphi(\xi)\,d\xi\right] + \frac{1}{2}\int_{x-t}^{x+t}\psi(\xi)\,d\xi \\ &= \frac{1}{2}(\varphi(x+t) + \varphi(x-t)) + \frac{1}{2}\int_{x-t}^{x+t}\psi(\xi)\,d\xi, \end{aligned}$$

and this is d'Alembert's solution.

Chapter 4

Dirichlet and Neumann Problems

4.1 Laplace's Equation and Harmonic Functions

2. Since $x^* = x - a$ and $y^* = y - b$, a straightforward chain rule differentiation suffices:

$$v_{x^*} = u_x x_{x^*} = u_x$$

and similarly,

$$v_{x^*x^*} = u_{xx}, v_{y^*} = u_y \text{ and } v_{y^*y^*} = u_{yy}.$$

If $u_{xx} + u_{yy} = 0$, then $v_{x^*x^*} + v_{y^*y^*} = 0$ as well

4. (a) $\mathbf{x}/|\mathbf{x}|$ is a unit vector in the direction from the origin to $\mathbf{x}$, and $I(\mathbf{x})$ is a positive scalar multiple of this, hence has the same direction. Further, the product of the distance from the origin to $\mathbf{x}$ and from the origin to $I(\mathbf{x})$ is

$$\begin{aligned} |I(\mathbf{x})||\mathbf{x}| &= \left|\frac{a^2}{|\mathbf{x}|^2}\mathbf{x}\right||\mathbf{x}| \\ &= \left(\frac{a^2}{|\mathbf{x}|^2}\right)|\mathbf{x}|^2 = a^2. \end{aligned}$$

(b) It is routine to verify that, for any $\mathbf{x} \neq \mathbf{0}$,

$$I(I(\mathbf{x})) = \mathbf{x}.$$

Further, if $|\mathbf{y}| > a$, then $|I(\mathbf{y})| < a$. Thus, let $\mathbf{x} = I(\mathbf{y})$, then we have $\mathbf{x}$ such that $I(\mathbf{x}) = \mathbf{y}$.

Solutions Manual to Accompany Beginning Partial Differential Equations, Third Edition. Peter V. O'Neil.

Further, such an $\mathbf{x}$ is uniquely determined by $\mathbf{y}$. For if

$$I(\mathbf{x}_1) = I(\mathbf{x}_2),$$

then by applying I again to both sides of this equation, we obtain $\mathbf{x}_1 = \mathbf{x}_2$. (c) is a straightforward exercise in chain rule differentiation.

4.2 The Dirichlet Problem for a Rectangle

2. Let $u(x,y) = X(x)Y(y)$ to obtain

$$X'' + \lambda X = 0,$$
$$Y'' - \lambda Y = 0,\ Y(0) = Y(2) = 0.$$

Solve for the eigenvalues and eigenfunctions of the problem for Y to get

$$\lambda_n = \frac{n^2\pi^2}{4},\ Y_n(y) = \sin(n\pi y/2).$$

Now the differential equation for X is

$$X'' - \frac{n^2\pi^2}{4}X = 0,$$

with solutions

$$X_n(x) = c_n e^{n\pi x/2} + d_n e^{-n\pi x/2}.$$

Since $u(3,y) = X(3)Y(y) = 0$, $X(3) = 0$,

$$c_n e^{3n\pi/2} + d_n e^{-3n\pi/2} = 0,$$

therefore

$$d_n = -c_n e^{3n\pi}.$$

This leads us to attempt a solution of the form

$$u(x,y) = \sum_{n=1}^{\infty} b_n \left(e^{n\pi x/2} - e^{3n\pi} e^{-n\pi x/2}\right) \sin(n\pi y/2).$$

We need

$$u(1,y) = y(2-y) = \sum_{n=1}^{\infty} b_n \left(1 - e^{3n\pi}\right) \sin(n\pi y/2).$$

This is like a Fourier series, and we can obtain the coefficients by reasoning as we did for Fourier series. Multiply both sides of this equation by $\sin(m\pi y/2)$ and integrate the resulting equation term by term to obtain

$$\int_0^2 \xi(2-\xi)\sin(m\pi\xi/2)\,d\xi$$
$$= \sum_{n=1}^{\infty} b_n(1 - e^{3n\pi}) \int_0^2 \sin(n\pi\xi/2)\sin(m\pi\xi/2)\,d\xi.$$

Two integrations yield

$$\int_0^2 \xi(2-\xi)\sin(m\pi\xi/2)\,d\xi = \frac{16(1-(-1)^m)}{m^3\pi^3}$$

and

$$\int_0^2 \sin(n\pi\xi/2)\sin(m\pi\xi/2)\,d\xi = \begin{cases} 0 & \text{if } n \neq m, \\ 1 & \text{if } n = m. \end{cases}$$

Therefore

$$b_n = \frac{16}{n^3\pi^3}\frac{1-(-1)^n}{1-e^{3n\pi}}.$$

4. Because $u(x,y)$ is nonzero on two sides of the domain square, define two problems. In problem 1,

$$u(0,y) = u(\pi,y) = u(x,\pi) = 0, u(x,0) = x(\pi - x)$$

while in problem 2,

$$u(x,0) = u(x,\pi) = u(\pi,y) = 0, u(0,y) = \sin(y).$$

Solve problem 1 by separation of variables to obtain the solution

$$u_1(x,y) = \sum_{n=1}^{\infty} \frac{4(1-(-1)^n)}{\pi n^3(1-e^{2n\pi})}\left(e^{ny} - e^{2n\pi}e^{-ny}\right)\sin(nx).$$

For problem 2, separation of variables leads to a solution of the form

$$u_2(x,y) = \sum_{n=1}^{\infty} c_n \left(e^{nx} - e^{2n\pi}e^{-nx}\right)\sin(ny).$$

To satisfy $u_2(0,y) = \sin(y)$, we must have

$$u_2(0,y) = \sum_{n=1}^{\infty} c_n(1 - 2^{2n\pi})\sin(ny).$$

This equation is satisfied if we choose $c_n = 0$ for $n = 2, 3, \ldots$, and $c_1 = 1/(1-e^{2\pi})$. Therefore

$$u_2(x,y) = \frac{1}{1-e^{2\pi}}\left(e^x - e^{2\pi}e^{-x}\right)\sin(y).$$

The solution of the problem is

$$u(x,y) = u_1(x,y) + u_2(x,y).$$

6. Let $u(x,y) = X(x)Y(y)$ to obtain

$$X'' + \lambda X = 0,\ X(0) = 0,$$
$$Y'' - \lambda Y = 0,\ Y(0) = Y'(K) = 0.$$

Begin with the problem for Y, which has eigenvalues and eigenfunctions

$$\lambda_n = -\left(\frac{(2n-1)\pi}{2K}\right)^2,\ Y_n(y) = \sin((2n-1)\pi y/2K).$$

With this solution for λ_n, the differential equation for X has solutions of the form

$$X_n(x) = \alpha_n e^{(2n-1)\pi x/2K} + \beta_n e^{-(2n-1)\pi x/2K}.$$

Because $X(0) = 0$, $\beta_n = -\alpha_n$ and $X_n(x)$ must be a constant multiple of $\sinh((2n-1)\pi x/2K)$. We are therefore led to propose a solution of the form

$$u(x,y) = \sum_{n=1}^{\infty} d_n \sinh\left(\frac{(2n-1)\pi}{2K}x\right)\sin\left(\frac{(2n-1)\pi}{2K}y\right).$$

To determine the coefficients, use the fact that

$$u(L,y) = g(y) = \sum_{n=1}^{\infty} \sinh\left(\frac{(2n-1)\pi L}{2K}\right)\sin\left(\frac{(2n-1)\pi}{2K}y\right).$$

Proceed as in problem 2 to find the coefficients by multiplying this equation by $\sin((2m-1)\pi y/2K)$ and integrating term by term. In the series, all terms with $n \neq m$ vanish, and we obtain

$$d_m = \frac{2}{K\sinh((2m-1)\pi L/2K)}\int_0^K g(\xi)\sinh\left(\frac{(2m-1)\pi}{2K}\xi\right)d\xi.$$

4.3 The Dirichlet Problem for a Disk

2. Compute the coefficients

$$a_0 = \frac{1}{\pi}\int_{-\pi}^{\pi}(\sin^3(\theta) + \cos^2(\theta))\,d\theta = 1,$$

$$a_n = \frac{1}{6^n\pi}\int_{-\pi}^{\pi}(\sin^3(\theta) + \cos^2(\theta))\cos(n\theta)\,d\theta$$

$$= \begin{cases} 0 & \text{if } n \neq 2, \\ 1/72 & \text{if } n = 2. \end{cases}$$

and

$$\begin{aligned} b_n &= \frac{1}{6^n \pi} \int_{-\pi}^{\pi} (\sin^3(\theta) + \cos^2(\theta)) \sin(n\theta)\, d\theta \\ &= \begin{cases} 0 & \text{if } n \neq 1, 3, \\ 3/4(6) & \text{if } n = 1, \\ -1/4(6^3) & \text{if } n = 3. \end{cases} \end{aligned}$$

We can write the solution compactly as

$$u(r, \theta) = \frac{1}{2} + \frac{1}{2}\left(\frac{r}{6}\right)^2 \cos(2\theta) + \frac{3}{4}\frac{r}{6}\sin(\theta) - \frac{1}{4}\left(\frac{r}{6}\right)^3 \sin(3\theta).$$

4. Convert the problem to polar coordinates, with $U(r, \theta) = u(r\cos(\theta), r\sin(\theta))$. The problem for $U(r, \theta)$ is

$$\begin{aligned} &U_{rr} + \frac{1}{r}U_r + \frac{1}{r^2}U_{\theta\theta} = 0 \\ &U(3, \theta) = 9\cos^2(\theta). \end{aligned}$$

It is routine to solve this problem to obtain

$$U(r, \theta) = \frac{9}{2} + \frac{1}{2}r^2\cos^2(\theta).$$

To write this solution in rectangular coordinates (in which the problem was stated), use the fact that

$$\cos(2\theta) = 2\cos^2(\theta) - 1 = 2\left(\frac{x}{r}\right)^2 - 1,$$

in which $r = \sqrt{x^2 + y^2}$. Therefore

$$u(x, y) = \frac{9}{2} + \frac{1}{2}\left(2\frac{x^2}{x^2 + y^2} - 1\right) = \frac{9}{2} + \frac{1}{2}(x^2 - y^2).$$

6. We know that the solution has the form

$$u(r, \theta) = \frac{1}{2}a_0 + \sum_{n=1}^{\infty}(a_n r^n \cos(n\theta) + b_n r^n \sin(n\theta)).$$

If $f(\theta)$ is an odd function of θ), then $f(\theta)\cos(n\theta)$ is odd, so $a_n = 0$ for $n = 0, 1, 2, \ldots$ and the solution is an odd function of θ.

If $f(\theta)$ is an even function of θ, then $f(\theta)\sin(n\theta)$ is an odd function and $b_n = 0$ for $n = 1, 2, \ldots$. Now the solution is even in θ.

7. Because the origin is not an element of the annulus, include terms involving r^{-n} in the solution, with n any positive integer. Also include a $\ln(r)$ term. Thus attempt a solution

$$u(r,\theta) = \frac{1}{2}a_0 + \sum_{n=1}^{\infty}(a_n r^n \cos(n\theta) + b_n r^n \sin(n\theta))$$
$$+ \sum_{n=1}^{\infty}(c_n r^{-n}\cos(n\theta) + c_n r^{-n}\sin(n\theta)) + k\ln(r).$$

From the boundary conditions,

$$u(\rho_1,\theta) = \frac{1}{2}a_0 + \sum_{n=1}^{\infty}(a_n\rho_1^n\cos(n\theta) + b_n\rho_1^n\sin(n\theta))$$
$$+ \sum_{n=1}^{\infty}(c_n\rho_1^{-n}\cos(n\theta) + d_n\rho_1^{-n}\sin(n\theta)) + k\ln(\rho_1 = g(\theta)$$

and

$$u(\rho_1,\theta) = \frac{1}{2}a_0 + \sum_{n=1}^{\infty}(a_n\rho_2^n\cos(n\theta) + b_n\rho_2^n\sin(n\theta))$$
$$+ \sum_{n=1}^{\infty}(c_n\rho_2^{-n}\cos(n\theta) + d_n\rho_2^{-n}\sin(n\theta)) + k\ln(\rho_2 = f(\theta)$$

Integrate these equations term by term from $-\pi$ to π. Most of the integrals are zero and we obtain

$$a_0\pi + 2k\pi\ln(\rho_1) = \int_{-\pi}^{\pi} g(\xi)\,d\xi$$

and

$$a_0\pi + 2k\pi\ln(\rho_2) = \int_{-\pi}^{\pi} f(\xi)\,d\xi.$$

Solve these equations for a_0 and k to get

$$k = \frac{1}{2\pi\ln(\rho_2/\rho_1)}\int_{-\pi}^{\pi}(f(\xi) - g(\xi))\,d\xi$$

and

$$a_0 = \frac{1}{\pi\ln(\rho_2/\rho_1)}\left(\ln(\rho_2)\int_{-\pi}^{\pi} g(\xi)\,d\xi - \ln(\rho_1)\int_{-\pi}^{\pi} f(\xi)\,d\xi\right).$$

To solve for the other coefficients, first multiply the series for $u(\rho_2,\theta)$ and $u(\rho_1,\theta)$ by $\cos(m\theta)$ and integrate from $-\pi$ to π. Most of the integrals vanish and we are left with

$$\pi\rho_1^m a_m + \pi\rho_1^{-m} c_m = \int_{-\pi}^{\pi} g(\xi)\cos(n\xi)\,d\xi,$$
$$\pi\rho_2^m a_m + \pi\rho_2^{-m} c_m = \int_{-\pi}^{\pi} f(\xi)\cos(m\xi)\,d\xi.$$

Solve these to obtain

$$a_m = \frac{1}{\pi(\rho_1^{2m} - \rho_2^{2m})}\left(\rho_1^m \int_{-\pi}^{\pi} g(\xi)\cos(m\xi)\,d\xi - \rho_2^m \int_{-\pi}^{\pi} f(\xi)\cos(m\xi)\,d\xi\right)$$

and

$$c_m = \frac{1}{\pi(\rho_2^{-2m} - \rho_1^{-2m})}\left(\rho_2^{-m} \int_{-\pi}^{\pi} f(\xi)\cos(m\xi)\,d\xi - \rho_1^{-m} \int_{-\pi}^{\pi} g(\xi)\cos(m\xi)\,d\xi\right)$$

for $m = 1, 2, \ldots$. For the remaining coefficients, multiply the series instead by $\sin(m\xi)$ and integrate to obtain

$$b_m = \frac{1}{\pi(\rho_1^{2m} - \rho_2^{2m})}\left(\rho_1^m \int_{-\pi}^{\pi} g(\xi)\sin(m\xi)\,d\xi - \rho_2^m \int_{-\pi}^{\pi} f(\xi)\sin(m\xi)\,d\xi\right)$$

and

$$d_m = \frac{1}{\pi(\rho_2^{-2m} - \rho_1^{-2m})}\left(\rho_2^{-m} \int_{-\pi}^{\pi} f(\xi)\sin(m\xi)\,d\xi - \rho_1^{-m} \int_{-\pi}^{\pi} g(\xi)\sin(m\xi)\,d\xi\right)$$

for $m = 1, 2, \ldots$.

8. Put $g(\theta) = 1$ and $f(\theta) = 2$ in the formulas for the coefficients derived for problem 7. Immediately we obtain

$$a_n = b_n = c_n = d_n = 0 \text{ for } n = 1, 2, \ldots.$$

Routine integrations also yield

$$a_0 = 2 \text{ and } k = \frac{1}{\ln(2)}.$$

The solution is

$$u(r,\theta) = 1 + \frac{1}{\ln(2)}\ln(r).$$

10. With $g(\theta) = \sin(\theta)$ and $f(\theta) = \cos(\theta)$, the coefficient formulas from problem 7 yield the solution

$$u(r,\theta) = \frac{2}{3}r\cos(\theta) - \frac{2}{3}\frac{1}{r}\cos(\theta) - \frac{1}{3}r\sin(\theta) + \frac{4}{3}\frac{1}{r}\sin(\theta).$$

12. Here $\rho_1 = 2$ and $\rho_2 = 4$, while $g(\theta) = \sin(2\theta)$ and $f(\theta) = \sin(4\theta)$. Now obtain the solution

$$\begin{aligned} u(r,\theta) &= -\frac{1}{60}r^2\sin(2\theta) + \frac{1}{255}r^4\sin(4\theta) \\ &\quad + \frac{64}{15}\frac{1}{r^2}\sin(2\theta) - \frac{256}{255}\frac{1}{r^4}\sin(4\theta). \end{aligned}$$

14. Let $R = \rho r$. This expands a disk of radius 1 to a disk of radius ρ. From a change of variables in equation 4.5, the solution on the disk of radius ρ is

$$u(R,\theta) = \frac{1}{2\pi}\int_{-\pi}^{\pi} \frac{1-(R/\rho)^2}{1-2(R/\rho)\cos(\theta-\xi)+(R/\rho)^2} f(\xi)\,d\xi.$$

After some algebraic manipulation, this yields equation 4.6.

16. First observe that $u(r,\theta) = r^n\sin(n\theta)$ is harmonic on the unit disk, for any positive integer n. Therefore u is the solution of the Dirichlet problem

$$\begin{aligned} \nabla^2 u(r,\theta) &= 0 \text{ for } 0 \le r < 1, \\ u(1,\theta) &= \sin(n\theta). \end{aligned}$$

But the solution of this problem is given by Poisson's integral formula, so

$$u(r,\theta) = r^n\sin(n\theta) = \frac{1}{2\pi}\int_{-\pi}^{\pi}\frac{1-r^2}{1-2r\cos(\theta-\xi)+r^2}\sin(n\xi)\,d\xi.$$

In particular, with $r = 1/2$ and $\theta = \pi/2$, we have

$$\begin{aligned} u(1/2,\pi/2) &= \frac{1}{2^n}\sin(n\pi/2) \\ &= \frac{1}{2\pi}\int_{-\pi}^{\pi}\frac{1-1/4}{1-\cos(\xi-\pi/2)+1/4}\sin(n\xi)\,d\xi \\ &= \frac{1}{2\pi}\int_{-\pi}^{\pi}\frac{3}{5-4\cos(\xi-\pi/2)}\sin(n\xi)\,d\xi. \end{aligned}$$

Since $\cos(\xi-\pi/2) = \sin(\xi)$, multiplication $2\pi/3$ yields

$$\int_{-\pi}^{\pi}\frac{1}{5-4\sin(\xi)}\sin(n\xi)\,d\xi = \frac{\pi}{3(2^{n-1})}\sin(n\pi/2).$$

17. The idea is the same as that used to solve problem 16. Write

$$r^n\cos(n\theta) = \frac{1}{2\pi}\int_{-\pi}^{\pi}\frac{1-r^2}{1-2r\cos(\theta-\xi)+r^2}\cos(n\xi)\,d\xi.$$

Then

$$u(1/2, \pi/2) = \frac{1}{2^n}\cos(n\pi/2) = \frac{1}{2\pi}\int_{-\pi}^{\pi} \frac{3}{5 - 4\cos(\xi - \pi/2)}\cos(n\xi/2)\,d\xi.$$

Then

$$\int_{-\pi}^{\pi} \frac{1}{5 - 4\sin(\xi)}\cos(n\xi)\,d\xi = \frac{\pi}{3(2^{n-1})}\cos(n\pi/2).$$

4.4 Properties of Harmonic Functions

4.4.1 Topology of R^n

2. Interior points are all points (x, y) with $x < 0, 1 < y < 4$. The boundary points are all points $(x, 1)$ and $(x, 4)$ with $x \leq 0$. The closure of P consists of all (x, y) with $x \leq 0$ and $1 \leq y \leq 4$. P is not open because P contains some boundary points (so not all points are interior points). P is not closed because there are boundary points of P that are not in P. P is connected.

4. W is an open, connected set, and is therefore also a domain. Boundary points are points $(-1, y)$ with $y \leq 1$, points $(6, y)$ with $y \leq 6$, and points (x, x) with $-1 \leq x \leq 6$. $\overline{W}$ consists of all points (x, y) with $-1 \leq x \leq 6$ and $x \leq y$. W is not closed because W does not contain all of its boundary points.

6. D has no interior points, and every point in R^2 is a boundary point. D is not open because it contains boundary points, and is not closed because there are boundary points not in the set. D is not connected. Finally, $\overline{D} = R^2$.

8. Geometrically, B consists of all points on or exterior to the ball of radius 2 about the origin in 3−space. The boundary points of B are points on the ball, that is, points (x, y, z) with $x^2 + y^2 + z^2 = 4$. Every point with $x^2 + y^2 + z^2 > 4$ (strictly outside the ball) is an interior point of B. B is not open because B contains boundary points. B is closed because it contains all of its boundary points. Further, $\overline{B} = B$. B is connected.

10. E consists of all points interior to the ball of radius 7 about $(-3, -4, -3)$, together with all points interior to the ball of radius 2 about $(11, 15, 12)$. Because the distance between the centers of these balls is

$$\sqrt{14^2 + 19^2 + 15^2},$$

which is approximately 27.96, these balls are disjoint, so E is not connected. Boundary points are points on the surfaces of the balls, and none of these is in E, so E is not closed. The points of E are all interior points, so E is open.

12. Suppose first that A is open. We want to show that A^C is closed. Suppose x is a boundary point of A^C. If $x \notin A^C$, then $x \in A$. But then x is an interior point, so there is an open ball B about x containing only points of A, and this contradicts x being a boundary point of A^C. There A^C must contain all of its boundary points (if there are any), and is therefore closed.

    Conversely, suppose A^C is closed. Let $x \in A$. We want to show that x must be an interior point of A, and that therefore A must be open. If every open ball B about x contains a point of A^C, then x would be a boundary point of A^C, and x would be in A^C because this set is closed. Therefore some open ball about x contains only points of A, so x is an interior point of A. Since every point of A is an interior point, A is open.

4.4.2 Representation Theorems

4. Let

$$u(\mathbf{y}) = \ln\left(\frac{1}{|\mathbf{y}-\mathbf{x}|}\right)$$

for $\mathbf{y} \in \Omega$ and $\mathbf{y} \neq \mathbf{x}$. We want to apply Green's second identity, but $\mathbf{v}(\mathbf{y}$ is not defined at $\mathbf{x}$. Place an open disk B of radius ϵ about $\mathbf{x}$, with ϵ chosen small enough that $B \subset \Omega$. Let Ω_ϵ be the set of points in the plane formed by removing all points of $\overline{B}$ from Ω. Then Ω_ϵ is a domain in R^2 and $\mathbf{v}$ is harmonic on Ω. By Green's second identity in the plane,

$$-\iint_{\Omega_\epsilon} \ln\left(\frac{1}{|\mathbf{y}-\mathbf{x}|}\right), dA_y$$
$$= \int_{\partial\Omega_\epsilon}\left[u\frac{\partial}{\partial n}\left(\ln\left(\frac{1}{|\mathbf{y}-\mathbf{x}|}\right)\right) - \ln\left(\frac{1}{|\mathbf{y}-\mathbf{x}|}\right)\frac{\partial u(\mathbf{y})}{\partial n}\right].$$

Now, $\partial\Omega_\epsilon$ consists of two disjoint sets. These are $\partial\Omega$ and the boundary circle C of B. The line integral in the last equation can therefore be written as a sum of the line integrals over these two disjoint paths, yielding

$$-\iint_{\overline{\Omega_\epsilon}} \ln\left(\frac{1}{|\mathbf{y}-\mathbf{x}|}\right)\nabla^2 u(\mathbf{y})\,dA_y$$
$$= \int_{\partial\Omega}\left[u\frac{\partial u}{\partial n}\left(\ln\left(\frac{1}{|\mathbf{y}-\mathbf{x}|}\right)\right) - \ln\left(\frac{1}{|\mathbf{y}-\mathbf{x}|}\right)\frac{\partial u(\mathbf{y})}{\partial n}\right] ds_y$$
$$+ \int_{C}\left[u\frac{\partial u}{\partial n}\left(\ln\left(\frac{1}{|\mathbf{y}-\mathbf{x}|}\right)\right) - \ln\left(\frac{1}{|\mathbf{y}-\mathbf{x}|}\right)\frac{\partial u(\mathbf{y})}{\partial n}\right] ds_y.$$

We want to take the limit in this equation as $\epsilon \to 0$. First, in this limit,

$$\iint_{\overline{\Omega_\epsilon}} \ln\left(\frac{1}{|\mathbf{y}-\mathbf{x}|}\right)\nabla^2 u(\mathbf{y})\,dA_y \to \iint_{\overline{\Omega}} \ln\left(\frac{1}{|\mathbf{y}-\mathbf{x}|}\right)\nabla^2 u(\mathbf{y})\,dA_y.$$

The line integral over $\partial\Omega$ is unaffected by the limit. This leads us to examine the line integral over C, which is a function of ϵ. If $\mathbf{y}$ is on C, then $|\mathbf{y}-\mathbf{x}| = \epsilon$. Now, the unit vector oriented along $-(\mathbf{y}-\mathbf{x})$ points into the disk (and out of Ω_ϵ). Thus let

$$\mathbf{n}(\mathbf{y}) = -\frac{\mathbf{y}-\mathbf{x}}{|\mathbf{y}-\mathbf{x}|} = -\frac{(y_1-x_1)\mathbf{i}+(y_2-x_2)\mathbf{j}}{\sqrt{(y_1-x_1)^2+(y_2-x_2)^2}}.$$

A straightforward computation gives us

$$\frac{\partial}{\partial n}\left(\ln\left(\frac{1}{|\mathbf{y}-\mathbf{x}|}\right)\right) = \nabla \ln\left(\frac{1}{|\mathbf{y}-\mathbf{x}|}\right)\cdot \mathbf{n}(\mathbf{y}) = \frac{1}{\epsilon}$$

for $\mathbf{y}$ on C. Therefore

$$\begin{aligned}
&\int_C \left[u\frac{\partial}{\partial n}\left(\ln\left(\frac{1}{|\mathbf{y}-\mathbf{x}|}\right)\right) - \ln\left(\frac{1}{|\mathbf{y}-\mathbf{x}|}\right)\frac{\partial u(\mathbf{y})}{\partial n}\right] ds_y \\
&= \int_C \left(\frac{1}{\epsilon}u(\mathbf{y}) - \ln(1/\epsilon)\frac{\partial u(\mathbf{y})}{\partial n}\right) ds_y \\
&= \frac{1}{\epsilon}\int_C u(\mathbf{x})\,ds_y + \int_C \left(\frac{1}{\epsilon}[u(\mathbf{y})-u(\mathbf{x})] - \ln(1/\epsilon)\frac{\partial u(\mathbf{y})}{\partial n}\right) ds_y.
\end{aligned}$$

Now,

$$\begin{aligned}
\frac{1}{\epsilon}\int_C u(\mathbf{x})\,ds_y &= \frac{1}{\epsilon}u(\mathbf{x})\int_C ds_y \\
&= \frac{1}{\epsilon}u(\mathbf{x})(2\pi\epsilon) = 2\pi u(\mathbf{x})
\end{aligned}$$

because $\int_C ds_y$ is the length of C. Finally,

$$\begin{aligned}
&\left|\int_C \left(\frac{1}{\epsilon}[u(\mathbf{y})-u(\mathbf{x})] - \ln(1/\epsilon)\frac{\partial u(\mathbf{y})}{\partial n}\right) ds_y\right| \\
&\leq \frac{1}{\epsilon}(2\pi\epsilon)\max_{\mathbf{y}\in C}|u(\mathbf{y})-u(\mathbf{x})| \\
&+ \ln(1/\epsilon)(2\pi\epsilon)\left|\frac{\partial u(\mathbf{y})}{\partial n}\right| \to 0
\end{aligned}$$

as $\epsilon \to 0$. Here we have used the continuity of u to conclude that $|u(\mathbf{y}) - u(\mathbf{x})| \to 0$ as $\mathbf{y} \to \mathbf{x}$, which occurs as $\epsilon \to 0$. Further, $|\partial u(\mathbf{y})/\partial n|$ is bounded and $\epsilon\ln(\epsilon) \to 0$ as $\epsilon \to 0$. Putting the pieces together, we obtain

$$\begin{aligned}
2\pi u(\mathbf{x}) = &\int_{\partial\Omega}\left(\ln\left(\frac{1}{|\mathbf{y}-\mathbf{x}|}\right)\frac{\partial u(\mathbf{y})}{\partial n} - u(\mathbf{y})\frac{\partial}{\partial n}\ln\left(\frac{1}{|\mathbf{y}-\mathbf{x}|}\right)\right) ds \\
&- \iint_\Omega \nabla^2 u(\mathbf{y})\ln\left(\frac{1}{|\mathbf{y}-\mathbf{x}|}\right) dA.
\end{aligned}$$

This yields the representation theorem in the plane.

4.4.3 The Mean Value Theorem and the Maximum Principle

2. Write

$$\begin{aligned}
\int_C u\frac{\partial v}{\partial n}\,dx &= \int_C u\left(\frac{\partial v}{\partial x}\mathbf{i}+\frac{\partial v}{\partial y}\mathbf{j}\right)\cdot\left(\frac{dy}{ds}\mathbf{i}-\frac{dx}{ds}\mathbf{j}\right)\,ds\\
&= \int_C -u\frac{\partial v}{\partial y}\,dx + u\frac{\partial v}{\partial x}\,dy\\
&= \iint_{\overline{\Omega}}\left[\frac{\partial}{\partial x}\left(u\frac{\partial v}{\partial x}\right)-\frac{\partial}{\partial y}\left(-u\frac{\partial v}{\partial y}\right)\right] \text{ by Green's theorem}\\
&= \iint_{\overline{\Omega}}\left[u\frac{\partial^2 v}{\partial x^2}+u\frac{\partial^2 v}{\partial y^2}+\frac{\partial u}{\partial x}\frac{\partial v}{\partial x}+\frac{\partial u}{\partial y}\frac{\partial v}{\partial y}\right]\,dA\\
&= \iint_{\overline{\Omega}}[u\nabla^2 v+\nabla u\cdot\nabla v]\,dA.
\end{aligned}$$

4. (a) Let $u = v$ in Green's first identity to write

$$\int_{\partial\Omega} u\frac{\partial u}{\partial n}\,ds = \iint_{\overline{\Omega}}(u\nabla^2 u+\nabla u\cdot\nabla u)\,dA.$$

But $\nabla^2 u = 0$ on Ω, so

$$\iint_{\overline{\Omega}} u\nabla^2 u\,dA = 0.$$

And $u = 0$ on $\partial\Omega$, so

$$\int_{\partial\Omega} u\frac{\partial u}{\partial n}\,ds = 0.$$

We conclude that

$$\iint_{\overline{\Omega}} |\nabla u|^2\,dA = 0.$$

Then $|\nabla u| = 0$, so

$$\frac{\partial u}{\partial x} = \frac{\partial u}{\partial y} = 0$$

on $\overline{\Omega}$. This means that $u(x,y)$ is constant on $\overline{\Omega}$. But $u = 0$ on $\partial\Omega$, so $u = 0$ on $\overline{\Omega}$.

(b) Here is another argument, using the maximum principle. Suppose that u is not constant on Ω. Then $u(x,y)$ achieves maximum and minimum values on $\overline{\Omega}$ only at boundary points of Ω. But $u(x,y) = 0$ on $\partial\Omega$, so the maximum and minimum values of $u(x,y)$ on $\overline{\Omega}$ are both zero, so these maximum and minimum values must be zero. This contradicts the assumption that u is not constant. We conclude that u must be a constant function on Ω. By continuity and the fact that $u(x,y) = 0$ on $\partial\Omega$, we must have $u(x,y) = 0$ on $\overline{\Omega}$.

6. Suppose u is harmonic and nonconstant on the entire plane. Use polar coordinates. By Harnack's inequality, for any positive number ρ, and $0 \leq r < \rho$,

$$\frac{\rho - r}{\rho + r} u(0,0) \leq u(r,\theta) \leq \frac{\rho + r}{\rho - r} u(0,0).$$

Because u is harmonic on the entire plane, ρ can be made as large as we want. Let $\rho \to \infty$ in Harnack's inequality to obtain

$$u(0,0) \leq u(r,\theta) \leq u(0,0),$$

which implies that $u(r,\theta) = u(0,0)$ for all points (r,θ).

4.5 The Neumann Problem

4.5.1 Uniqueness and Existence

2. Let u and v be solutions and let $w = u - v$. Then $\nabla^2 w = 0$ on Ω and

$$\begin{aligned} &\frac{\partial w}{\partial n} + hw \\ &= \frac{\partial u}{\partial n} + hu - \frac{\partial v}{\partial n} - hv = f - f = 0 \end{aligned}$$

on $\partial\Omega$. Apply Green's first identity in the plane to write

$$\int_{\partial\Omega} w \frac{\partial w}{\partial n}\, ds = \iint_{\Omega} (w\nabla^2 w + |\nabla w|^2)\, dA.$$

But $\nabla^2 w = 0$ on Ω and $\partial w/\partial n + hw = 0$ on $\partial\Omega$, so this equation becomes

$$\int_{\partial\Omega} -hw^2\, ds = \iint_{\Omega} |\nabla w|^2\, dA.$$

Therefore

$$\iint_{\Omega} |\nabla w|^2\, dA \leq 0.$$

This implies that $|\nabla w| = 0$ on Ω. But then $w_x = w_y = 0$ so w is constant on Ω. Then

$$\iint_{\Omega} |\nabla w|^2\, dA = 0,$$

so

$$\int_{\partial\Omega} -hw^2\, ds = 0.$$

Because h is nonnegative on $\partial\Omega$ and not identically zero, we must have $w^2 = 0$. Since w is constant on Ω, w must be identically zero, so $u = v$.

4. Suppose u and v are solutions and let $w = u - v$. Then $\nabla^2 w = -kw$ on Ω and $\partial w/\partial n = 0$ on $\partial\Omega$. Use Green's first identity to write

$$\int_{\partial\Omega} w\frac{\partial w}{\partial n}\, ds = 0 = \iint_{\Omega} (-kw^2 + |\nabla w|^2)\, dA.$$

Now $-k > 0$, so the integrand on the right cannot be negative:

$$-kw^2 + |\nabla w|^2 \geq 0$$

on Ω. Therefore, vanishing of the last double integral implies that

$$-kw^2 + |\nabla w|^2 = 0$$

on Ω, and therefore $w = 0$ on Ω.

4.5.2 Neumann Problem for a Rectangle

2. The solution is the sum of solutions to two problems, in each of which the boundary data is nonzero on only one side of the rectangle. First consider problem 1:

$$\begin{aligned}\nabla^2 u &= 0 \text{ for } 0 < x < 1, 0 < y < \pi,\\ u_x(0,y) &= y - \pi/2, u_x(1,y) = 0 \text{ for } 0 < y < \pi,\\ u_y(x,0) &= u_y(x,\pi) = 0 \text{ for } 0 < x < 1.\end{aligned}$$

First observe that the condition

$$\int_0^{\pi} \left(y - \frac{\pi}{2}\right) dy = 0,$$

which is necessary for a solution to exist, is satisfied. Use separation of variables to conclude that there is a solution of the form

$$u(x,y) = \alpha_0 + \sum_{n=1}^{\infty} a_n(e^{nx} + e^{2n}e^{-nx})\cos(ny).$$

Now

$$u_x(x,y) = \sum_{n=1}^{\infty} na_n(e^{nx} - e^{2n}e^{-nx})\cos(ny),$$

so

$$u_x(0,y) = y - \frac{\pi}{2} = \sum_{n=1}^{\infty}(1 - e^{2n})\cos(ny).$$

This is a Fourier cosine expansion on $[0,\pi]$, so choose

$$\begin{aligned}na_n(1 - e^{2n}) &= \frac{2}{\pi}\int_0^{\pi}\left(y - \frac{\pi}{2}\right)\cos(ny)\, dy\\ &= \frac{2}{\pi n^2}((-1)^n - 1).\end{aligned}$$

This determines the coefficients a_n and hence the solution of problem 1.

Now consider problem 2:

$$\begin{aligned}\nabla^2 u = 0 \text{ for } 0 < x < 1, 0 < y < \pi,\\ u_x(0,y) = 0, u_x(1,y) = \cos(y) \text{ for } 0 < y < \pi,\\ u_y(x,0) = u_y(x,\pi) = 0 \text{ for } 0 < x < 1.\end{aligned}$$

Separation of variables leads to a solution of the form

$$u(x,y) = \beta_0 + \sum_{n=1}^{\infty} b_n \cosh(nx)\cos(ny).$$

Because

$$u_x(0,y) = \sum_{n=1}^{\infty} nb_n \sinh(nx)\cos(ny),$$

we need

$$u_x(1,y) = \sum_{n=1}^{\infty} nb_n \sinh(n)\cos(ny) = \cos(y).$$

By inspection, choose $b_n = 0$ for $n = 2, 3, \ldots$ and $b_1 \sinh(1) = 1$. The solution of problem 2 is

$$u_2(x,y) = \beta_0 + \frac{1}{\sinh(1)}\cosh(x)\cos(y).$$

The solution of the given problem is the sum of the solutions of problems 1 and 2.

4. Separation of variables leads to the solution

$$u(x,t) =$$
$$\sum_{n=1}^{\infty} \frac{2}{\pi(1-e^{2n\pi})}\left(\int_0^{\pi} f(\xi)\cos(n\xi)\,d\xi\right)(e^{ny} - e^{2n\pi}e^{-ny})\cos(nx).$$

4.5.3 Neumann Problem for a Disk

2. The solution has the form

$$u(r,\theta) =$$
$$\frac{1}{2}a_0 + \frac{\rho}{\pi}\sum_{n=1}^{\infty}\left(\frac{r}{\rho}\right)^n \int_{-\pi}^{\pi} (\cos(n\xi)\cos(n\theta) + \sin(n\xi)\sin(n\theta))\cos(3\xi)\,d\xi.$$

Compute

$$\int_{-\pi}^{\pi} \sin(n\xi)\cos(3\xi)\,d\xi = 0$$

for $n = 1, 2, 4, \ldots$. And

$$\int_{-\pi}^{\pi} \cos(n\xi)\cos(3\xi)\,d\xi = \begin{cases} 0 & \text{if } n = 1, 2, 4, 5, \cdots, \\ \pi & \text{if } n = 3. \end{cases}$$

Therefore the solution is

$$\begin{aligned} u(r,\theta) &= \alpha_0 + \frac{\rho}{2\pi}\left(\frac{r}{\rho}\right)^2 \pi\cos(3\theta) \\ &= \alpha_0 + \frac{1}{2\rho}r^2\cos(3\theta), \end{aligned}$$

in which α_0 is an arbitrary constant.

4. In polar coordinates the boundary condition is

$$\frac{\partial u}{\partial r}(1,\theta) = \cos(\theta).$$

The solution is

$$u(r,\theta) = \alpha_0 + \frac{1}{\pi} r\cos(\theta) \int_{-\pi}^{\pi} \cos^2(\xi)\,d\xi = r\cos(\theta).$$

In rectangular coordinates, the solution is $u(x, y) = x$.

4.6 Poisson's Equation

2. Write

$$F(x,y) = \sum_{n=1}^{\infty} f_n(y)\sin(nx) = -14\sin(3x) + \sqrt{2}\sin(12x).$$

Then

$$f_3(y) = -14, f_{13}(y) = \sqrt{2} \text{ and } f_n(y) = 0 \text{ for } n \neq 3, 12.$$

First solve

$$Y_3''(y) - 9Y_3(y) = f_3(y); Y_3(0) = Y_3(4) = 0.$$

We obtain

$$\begin{aligned} Y_3(y) = &\frac{7}{9\sinh(12)}(e^{-12} - 1)e^{3y} \\ &+ \frac{7}{9\sinh(12)}(1 - e^{12})e^{-3y} + \frac{14}{9}. \end{aligned}$$

In similar fashion, obtain

$$Y_{12}(y) = \frac{\sqrt{2}}{288\sinh(48)}(1 - e^{-48})e^{12y} + \frac{\sqrt{2}}{288\sinh(48)}(e^{48} - 1)e^{-12y} - \frac{\sqrt{2}}{144}.$$

The solution is

$$u(x,y) = Y_3(y)\sin(3x) + Y_{12}(y)\sin(12x).$$

4. Write

$$F(x,y) = \sqrt{3}y = \sum_{n=1}^{\infty} f_n(y)\sin(nx),$$

in which

$$f_n(y) = \frac{2}{\pi}\int_0^{\pi} \sqrt{3}\xi \sin(n\xi)\,d\xi = \frac{2\sqrt{3}}{n\pi}(1 - (-1)^n)y.$$

Solve

$$Y_n'' - n^2 Y_n = f_n(y);\; Y_n(0) = Y_n(2\pi) = 0.$$

Since $f_n(y) = 0$ if n is even, then $Y_n(y) = 0$ for even n. If n is odd we obtain

$$Y_n(y) = \frac{2\pi k_n}{n^2\sinh(2\pi n)}\sinh(ny) - \frac{k_n}{n^2}y,$$

where

$$k_n = \frac{2\sqrt{3}}{n\pi}(1 - (-1)^n) = \frac{4\sqrt{3}}{n\pi},$$

because n is odd. The solution is

$$u(x,y) = \sum_{n=1}^{\infty} Y_{2n-1}(y)\sin((2n-1)x).$$

4.7 An Existence Theorem for the Dirichlet Problem

2. If P is in D and not in K, then

$$f_{K,v_1+v_2}(P) = v_1(P) + v_2(P) = f_{K,v_1}(P) + f_{K,v_2}(P).$$

For points in K, recall that f_{K,v_1} is that unique function that is harmonic on K and equal to v_1 on ∂K, while f_{K,v_2} is that unique function that is harmonic on K and equal to v_2 on ∂K. Then f_{K,v_1+v_2} is that unique function that is harmonic on K and equal to $v_1 + v_2$ on ∂K. Therefore, on K,

$$f_{K,v_1+v_2} = f_{K,v_1} + f_{K,v_2}.$$

Chapter 5

Fourier Integral Methods of Solution

5.1 The Fourier Integral of a Function

2. The Fourier integral of $f(x)$ is

$$\int_0^\infty \frac{2}{\pi}\frac{\sin(\alpha)\cos(\alpha\omega)+\omega\cos(\alpha)\sin(\alpha\omega)}{\omega^2-1}\cos(\omega x)\,d\omega.$$

This converges to

$$\begin{cases}\cos(x) & \text{for } -\alpha < x < \alpha,\\ \cos(\alpha)/2 & \text{for } x = \pm\alpha,\\ 0 & \text{for } |x| > \alpha.\end{cases}$$

4.

$$\int_0^\infty \frac{4}{\pi}\frac{\omega}{(1+\omega^2)^2}\sin(\omega x)\,d\omega,$$

converging to $f(x)$ for all x.

6.

$$\int_0^\infty \frac{2k}{\pi}\frac{1-\cos(\alpha\omega)}{\omega}\sin(\alpha\omega)\,d\omega,$$

Solutions Manual to Accompany Beginning Partial Differential Equations, Third Edition. Peter V. O'Neil.

converging to

$$\begin{cases} k & \text{for } 0 < x < \alpha, \\ -k & \text{for } -\alpha < x < 0, \\ k/2 & \text{for } x = \alpha, \\ -k/2 & \text{for } x = -\alpha, \\ 0 & \text{for } |x| > \alpha \end{cases}$$

10. The sine integral representation of $f(x)$ is

$$\frac{2}{\pi}\int_0^\infty \frac{\omega}{\omega^2 - \pi^2}(1 + \cos(\omega))\sin(\omega x)\,d\omega.$$

This integral converges to

$$\begin{cases} \cos(\pi x) & \text{for } 0 < x < 1, \\ 0 & \text{for } = 0 \text{ and for } x \geq 1. \end{cases}$$

The cosine integral is

$$\frac{2}{\pi}\int_0^\infty \frac{\omega}{\omega^2 - \pi^2}\sin(\omega)\cos(\omega x)\,d\omega,$$

converging to

$$\begin{cases} \cos(\pi x) & \text{for } 0 \leq x < 1, \\ -1/2 & \text{for } x = 1, \\ 0 & \text{for } x > 1. \end{cases}$$

12. The sine integral is

$$\frac{12}{\pi}\int_0^\infty \frac{\omega}{(9 + \omega^2)^2}\sin(\omega x)\,d\omega,$$

converging to xe^{-3x} for $x \geq 0$ (note that $f(0) = 0$).
The cosine integral is

$$\frac{9}{\pi}\int_0^\infty \frac{2 - \omega^2}{(9 + \omega^2)^2}\cos(\omega x)\,d\omega,$$

converging to xe^{-3x} for $x \geq 0$.

14. The sine integral is

$$\frac{2}{\pi}\int_0^\infty \frac{\omega^2\alpha^2\cos(\alpha\omega) - 2\cos(\alpha\omega) - 2\alpha\omega\sin(\alpha\omega)}{\omega^3}\sin(\omega x)\,d\omega,$$

converging to

$$\begin{cases} x^2 & \text{for } 0 < x < \alpha, \\ \alpha^2/2 & \text{for } x = \alpha, \\ 0 & \text{for } x = 0 \text{ and for } x > \alpha. \end{cases}$$

The cosine integral is

$$\frac{2}{\pi}\int_0^\infty \frac{\omega^2\alpha^2\sin(\alpha\omega)-2\sin(\alpha\omega)+2\alpha\omega\cos(\alpha\omega)}{\omega^3}\cos(\omega x)\,d\omega,$$

converging to

$$\begin{cases} x^2 & \text{for } 0 \le x < \alpha, \\ \alpha^2/2 & \text{for } x = \alpha, \\ 0 & \text{for } x > \alpha. \end{cases}$$

15. The cosine integral expansion of $g(x)$ is

$$\int_0^\infty a_\omega \cos(\omega x)\,d\omega,$$

where

$$a_\omega = \frac{2}{\pi}\int_0^\infty \frac{1}{1+\xi^2}\cos(\omega\xi)\,d\xi = e^{-\omega}$$

by the first of the Laplace integrals.

The sine expansion of $h(x)$ is

$$\int_0^\infty b_\omega \sin(\omega x)\,d\omega,$$

where

$$b_\omega = \frac{2}{\pi}\int_0^\infty h(\xi)\sin(\omega\xi)\,d\xi = \begin{cases} e^{-\omega} & \text{for } \omega > 0, \\ 0 & \text{for } \omega = 0. \end{cases}$$

16. The Fourier cosine integral of $g''(x)$ has the form

$$\int_0^\infty a_\omega \cos(\omega x)\,d\omega$$

Write the integral formula for a_ω and integrate twice by parts, making use of the assumptions about limits of $g(x)$ and $g'(x)$ as $x \to \infty$, and also the fact that $g'(0) = 0$. We obtain

$$\begin{aligned} a_\omega &= \frac{2}{\pi}\int_0^\infty g''(\xi)\cos(\omega\xi)\,d\xi \\ &= \left[\frac{2}{\pi}g'(\xi)\cos(\omega\xi)\right]_0^\infty + \frac{2\omega}{\pi}\int_0^\infty g'(\xi)\sin(\omega\xi)\,d\xi \\ &= \frac{2\omega}{\pi}\left[g(\xi)\sin(\omega\xi)\right]_0^\infty - \frac{2\omega}{\pi}\int_0^\infty g(\xi)\omega\cos(\omega\xi)\,d\xi \\ &= -\frac{2\omega^2}{\pi}\int_0^\infty g(\xi)\cos(\omega\xi)\,d\xi \end{aligned}$$

and this is $-2\omega^2/\pi$ times the Fourier cosine integral coefficient of $g(x)$.

17. Compute

$$a'(\omega) = -\int_{-\infty}^{\infty} tg(t)\sin(\omega t)\,dt$$

and

$$a''(\omega) = -\int_{-\infty}^{\infty} t^2 g(t)\cos(\omega t)\,dt.$$

Now recognize these as $\pi/2$ times the Fourier cosine integral coefficients of $xh(x)$ and $x^2h(x)$, respectively.

5.2 The Heat Equation on the Real Line

2. The solution is

$$\begin{aligned} u(x,t) &= \frac{1}{2\sqrt{\pi kt}} \int_{-\infty}^{\infty} e^{-(x-\xi)^2/4kt} f(\xi)\,d\xi \\ &= \frac{1}{2\sqrt{\pi kt}} \int_{-a}^{a} e^{-(x-\xi)^2/4kt} f(\xi)\,d\xi \end{aligned}$$

and this is positive for all t if $f(\xi) > 0$ for $-a < \xi < a$.

4. The solution by Fourier integral is

$$u(x,t) = \int_0^{\infty} (a_\omega \cos(\omega x) + b_\omega \sin(\omega x))e^{-\omega^2 kt}\,d\omega,$$

in which

$$a_\omega = \frac{1}{\pi}\int_{-\pi}^{\pi} \sin(\xi)\cos(\omega\xi)\,d\xi = 0$$

and, if $\omega \neq 1$,

$$b_\omega = \frac{1}{\pi}\int_{-\pi}^{\pi} \sin(\xi)\sin(\omega\xi)\,d\xi = \frac{2}{\pi}\frac{\sin(\pi\omega)}{1-\omega^2}.$$

If $\omega = 1$, we obtain

$$b_1 = \frac{1}{\pi}\int_{-\pi}^{\pi} \sin^2(\xi)\,d\xi = 1,$$

which is the limit of $2\sin(\pi\omega)/(1-\omega^2)$ as $\omega \to 1$. Therefore the solution is

$$u(x,t) = \frac{2}{\pi}\int_0^{\infty} \frac{\sin(\pi\omega)}{1-\omega^2}\sin(\omega x)e^{-k\omega^2 t}\,d\omega.$$

Alternatively, we may use equation 5.9 to write the solution as

$$u(x,t) = \frac{1}{2\sqrt{\pi kt}}\int_{-\pi}^{\pi} e^{-(x-\xi)^2/4kt}\sin(\xi)\,d\xi.$$

6. The solution by Fourier integral is

$$u(x,t) = \int_0^\infty (a_\omega \cos(\omega x) + b_\omega \sin(\omega x))e^{-\omega^2 kt}\, d\omega$$

where

$$\begin{aligned} a_\omega &= \frac{1}{\pi}\int_{-1}^{1} e^{-\xi}\cos(\omega\xi)\, d\xi \\ &= \frac{1}{\pi(1+\omega^2)}(-\cos(\omega) + \omega\sin(\omega) + e^2\cos(\omega) + e^2\omega\sin(\omega)) \end{aligned}$$

and

$$\begin{aligned} b_\omega &= \frac{1}{\pi}\int_{-1}^{1} e^{-\xi}\sin(\omega\xi)\, d\xi \\ &= \frac{1}{\pi(1+\omega^2)}(-\omega\cos(\omega) - \sin(\omega) + e^2\omega\cos(\omega) - e^2\sin(\omega)). \end{aligned}$$

We can also write an integral solution using equation 5.9:

$$u(x,t) = \frac{2}{2\sqrt{\pi kt}}\int_{-1}^{1} e^{-\xi}e^{-(x-\xi)^2/4kt}\, d\xi.$$

8. Use equation 5.9 to manipulate the solution as follows:

$$\begin{aligned} u(x,t) &= \frac{}{2\sqrt{\pi kt}}\int_{-1}^{1} e^{-(x-\xi)^2/4kt}\, d\xi \\ &= \frac{1}{2\sqrt{\pi kt}}\int_{-1}^{1}\sum_{n=0}^{\infty}\frac{1}{n!}\left(\frac{(x-\xi)^2}{4kt}\right)^n d\xi \\ &= \frac{2}{2\sqrt{\pi kt}}\sum_{n=0}^{\infty}\frac{1}{n!}\frac{-1)^n}{(4kt)^n}\int_{-1}^{1}(x-\xi)^{2n}\, d\xi \\ &= \frac{1}{2\sqrt{\pi kt}}\sum_{n=0}^{\infty}\frac{(-1)^n}{(2n+1)(4kt)^n}\left[(x+1)^{2n+1} - (x-1)^{2n+1}\right]. \end{aligned}$$

9. Proceed by a routine separation of variables to obtain a solution of the form

$$u(x,t) = \int_0^\infty (a_\omega\cos(\omega x) + b_\omega\sin(\omega x))e^{-\omega^2 kt}\, d\omega.$$

To satisfy the initial condition, we need

$$\begin{aligned} u_x(x,0) &= \int_0^\infty (-\omega a_\omega \sin(\omega x) + \omega b_\omega\cos(\omega x))\, d\omega \\ &= g(x) = \begin{cases} \cos(\pi x) & \text{for } -1 \le x \le 1, \\ 0 & \text{for } |x| > 1. \end{cases} \end{aligned}$$

This is a Fourier integral expansion of $g(x)$, so choose

$$\omega b_\omega = \frac{1}{\pi}\int_{-1}^{1} \cos(\pi\xi)\cos(\omega\xi)\,d\xi$$

and

$$-\omega a_\omega = \frac{1}{\pi}\int_{-1}^{1} \cos(\pi\xi)\sin(\omega\xi)\,d\xi.$$

Compute

$$b_\omega = \begin{cases} 2\sin(\omega)/\pi(\pi^2-\omega^2) & \text{for } \omega \neq \pi, \\ 1/\pi(\pi^2-\omega^2) & \text{for } \omega = \pi \end{cases}$$

and

$$a_\omega = 0 \text{ for all } \omega \geq 0.$$

In this expression for b_ω, notice that, in the limit as $\omega \to \pi$,

$$\frac{2\sin(\omega)}{\pi(\pi^2-\omega^2)} \to \frac{1}{\pi^2}.$$

The solution is

$$u(x,t) = \int_0^\infty \frac{2\sin(\omega)}{\pi(\pi^2-\omega^2)}\sin(\omega x)e^{-\omega^2 kt}\,d\omega.$$

12. By separation of variables we find the solution

$$u(x,t) = \frac{2}{\pi}\int_0^\infty \frac{\omega}{\alpha^2+\omega^2}\sin(\omega x)e^{-k\omega^2 t}\,d\omega.$$

We can also write a different integral formula for the solution:

$$u(x,t) = \frac{1}{2\sqrt{\pi kt}}\int_0^\infty \left(e^{-(x-\xi)^2/4kt} - e^{-(x+\xi)^2/4kt}\right)e^{-\alpha\xi}\,d\xi.$$

14. By separation of variables,

$$u(x,t) = \frac{2}{\pi}\int_0^\infty \frac{1-\cos(h\omega)}{\omega}\sin(\omega x)e^{-k\omega^2 t}\,d\omega.$$

Another integral formulation of the solution is

$$u(x,t) = \frac{1}{2\sqrt{\pi kt}}\int_0^h \left(e^{-(x-\xi)^2/4kt} - e^{-(x+\xi)^2/4kt}\right)d\xi.$$

16. Using separation of variables, obtain

$$u(x,t) = \frac{2}{\pi}\int_0^\infty \frac{1}{\omega(1-2\cos(h\omega)+\cos(2h\omega))}e^{-k\omega^2 t}\,d\omega.$$

We can also write a solution

$$u(x,t) = \frac{1}{2\sqrt{\pi kt}}\int_0^h \left(e^{-(x-\xi)^2/4kt} - e^{-(x+\xi)^2/4kt}\right)d\xi$$
$$- \frac{1}{2\sqrt{\pi kt}}\int_h^{2h} \left(e^{-(x-\xi)^2/4kt} - e^{-(x+\xi)^2/4kt}\right)d\xi.$$

5.3 The Debate Over the Age of the Earth

5.4 Burgers' Equation

2. First compute

$$g(x) = e^{\frac{1}{2k}\int_0^x f(\xi)\,d\xi}.$$

With the given $f(x)$, we find that

$$g(x) = \begin{cases} e^{-x/2k} & \text{for } -1 \le x \le 1, \\ e^{-1/2k} & \text{for } x > 1, \\ e^{1/2k} & \text{for } x < -1. \end{cases}$$

The solution is

$$\begin{aligned} u(x,t) = & \frac{\int_{-1}^{1} \frac{x-\xi}{t} e^{-\xi/2k} e^{-(x-\xi)^2/4kt}}{\int_{-1}^{1} e^{-\xi/2k} e^{-(x-\xi)^2/4kt}} \\ & + \frac{\int_1^\infty \frac{x-\xi}{t} e^{-1/2k} e^{-(x-\xi)^2/4kt}\,d\xi}{\int_1^\infty e^{-1/2k} e^{-(x-\xi)^2/4kt}\,d\xi} \\ & + \frac{\int_{-\infty}^{-1} \frac{x-\xi}{t} e^{1/2k} e^{-(x-\xi)^2/4kt}\,d\xi}{\int_{-\infty}^{-1} e^{1/2k} e^{-(x-\xi)^2/4kt}\,d\xi}. \end{aligned}$$

4. The proposition follows easily from recognizing that

$$e^{-H(\xi,x,t)/2k} = e^{-(x-\xi)^2/4kt} e^{-(1/2k)\int_0^\xi f(w)\,dw} = g(\xi) e^{-(x-\xi)^2/4kt}.$$

5. Compute

$$\begin{aligned} E'(t) &= \int_0^1 2uu_t\,dx \\ &= \int_0^1 2u(ku_{xx} - uu_x)\,dx \\ &= 2k\int_0^1 uu_{xx}\,dx - 2\int_0^1 u^2u_x\,dx \\ &= 2k\int_0^1 uu_{xx}\,dx - 2\int_0^1 \frac{\partial}{\partial x}\left(\frac{1}{3}u^3\right)dx \\ &= 2k\int_0^1 uu_{xx}\,dx - \frac{2}{3}\left(u(1,t)^3 - u(0,t)^3\right). \end{aligned}$$

Since $u(0,t) = u(1,t) = 0$, we now have an equation involving only one integral, which we can integrate by parts:

$$\begin{aligned} E'(t) &= 2k \int_0^1 u u_{xx}\, dx \\ &= 2k \left[u u_x\right]_0^1 - 2k \int_0^1 u_x^2\, dx \\ &= -2k \int_0^1 u_x^2\, dx. \end{aligned}$$

Here we again used the fact that $u(1,t) = u(0,t) = 0$. We therefore have

$$E'(t) \leq 0,$$

so $E(t)$ is nonincreasing. But

$$E(0) = \int_0^1 u(x,0)^2\, dx = \int_0^1 f(\xi)^2\, d\xi.$$

Therefore

$$0 \leq E(t) \leq \int_0^1 f(\xi)^2\, d\xi.$$

8. First compute

$$c = \frac{1}{2}(L_+ + L_-) = 12.$$

The traveling wave solution is

$$u(x,t) = \frac{20 + 4e^{8(x-12t)/5}}{1 + e^{8(x-12t)/5}}.$$

5.5 The Cauchy Problem for the Wave Equation

2. The solution is

$$u(x,t) = \frac{4}{\pi} \int_0^\infty \frac{\omega \cos(\omega) - \sin(\omega)}{\omega^3} \cos(\omega x) \cos(\omega c t)\, d\omega.$$

4. The solution is

$$u(x,t) = \frac{16}{\pi} \int_0^\infty \cos^2(\omega) \frac{\sin^2(\omega)}{\omega^2} \cos(\omega x) \cos(\omega c t)\, d\omega.$$

6. The solution is

$$u(x,t) = \frac{16}{\pi} \int_0^\infty \sin(\pi \omega) \frac{1 - 2\cos^2(\omega)}{\omega(\omega^2 - 4)} \cos(\omega x) \cos(\omega c t)\, d\omega.$$

7. We want to solve the problem

$$u_{tt} = c^2 u_{xx},$$
$$u(x,0) = 0, u_t(x,0) = \psi(x)$$

with $-\infty < x < \infty$.

Let $u(x,t) = X(x)T(t)$ to separate the variables. As with the case of zero initial velocity, obtain eigenvalues and eigenfunctions

$$\lambda = \omega^2,\ X_\omega(x) = a_\omega \cos(\omega x) + b_\omega \sin(\omega x).$$

The problem for T is

$$T'' + \omega^2 c^2 T = 0,\ T(0) = 0$$

with solutions

$$T_\omega(t) = \sin(\omega c t).$$

Thus obtain functions

$$u_\omega(x,t) = \int_0^\infty (a_\omega \cos(\omega x) + b_\omega \sin(\omega x)) \sin(\omega c t)\, d\omega.$$

To solve for the coefficients, use the initial velocity condition:

$$u_t(x,0) = \int_0^\infty (c\omega a_\omega \cos(\omega x) + c\omega b_\omega \sin(\omega x))\, d\omega = \psi(x).$$

This is a Fourier integral expansion of the initial velocity function, with coefficients $c\omega a_\omega$ and $c_\omega b_\omega$. Thus choose

$$a_\omega = \frac{1}{\pi\omega c} \int_{-\infty}^{\infty} \psi(\xi) \cos(\omega\xi)\, d\xi$$

and

$$b_\omega = \frac{1}{\pi\omega c} \int_{-\infty}^{\infty} \psi(\xi) \sin(\omega\xi)\, d\xi.$$

8. The solution is

$$u(x,t) = \int_0^\infty \frac{2}{\pi\omega c(1+\omega^2)} \cos(\omega x) \sin(\omega c t)\, d\omega.$$

10. The solution is

$$u(x,t) = \int_0^\infty \frac{2}{\pi\omega c} \frac{\cos(\pi\omega/2)}{1-\omega^2} \cos(\omega x) \sin(\omega c t)\, d\omega.$$

12. The solution is

$$u(x,t) = \int_0^\infty \frac{12\omega}{\pi(9+\omega^2)} \sin(\omega x) \cos(\omega c t)\, d\omega.$$

14. The solution is

$$u(x,t) = \int_0^\infty b_\omega \sin(\omega x)\sin(\omega ct)\,d\omega,$$

where

$$\begin{aligned} b_\omega &= \frac{2}{\pi\omega c}\int_0^\infty \psi(\xi)\sin(\omega\xi)\,d\xi \\ &= \frac{2}{\pi\omega c}\left[\frac{K}{\omega}(1-\cos(K\omega)) + \frac{1}{\omega^2}(\sin(K\omega) - \sin(2K\omega) + K\omega\cos(2K\omega))\right]. \end{aligned}$$

5.6 Laplace's Equation on Unbounded Domains

2. The solution is

$$\begin{aligned} u(x,y) &= \frac{y}{\pi}\int_{-\infty}^\infty \frac{1}{y^2+(\xi-x)^2} f(\xi)\,d\xi \\ &= \frac{Ky}{\pi}\int_0^\alpha \frac{1}{y^2+(\xi-x)^2}\,d\xi \\ &= \frac{K}{\pi}\left(\arctan\left(\frac{x}{y}\right) + \arctan\left(\frac{\alpha-x}{y}\right)\right). \end{aligned}$$

4. The solution is

$$\begin{aligned} u(x,y) &= \frac{y}{\pi}\left[\int_{-1}^\alpha \frac{1}{y^2+(\xi-x)^2}\,d\xi + \int_\alpha^{2\alpha} \frac{2}{y^2+(\xi-x)^2}\right] \\ &= \frac{1}{\pi}\left[\arctan\left(\frac{1+x}{y}\right) - \arctan\left(\frac{\alpha-x}{y}\right) + 2\arctan\left(\frac{2\alpha-x}{y}\right)\right]. \end{aligned}$$

6. The solution is

$$\begin{aligned} u(x,y) &= \frac{y}{\pi}\int_0^3 \left(\frac{1}{y^2+(\xi-x)^2} - \frac{1}{y^2+(\xi+x)^2}\right)d\xi \\ &\quad - \frac{y}{\pi}\int_3^5 \left(\frac{1}{y^2+(\xi-x)^2} - \frac{1}{y^2+(\xi+x)^2}\right)d\xi \\ &= \frac{1}{\pi}\left[2\arctan\left(\frac{x}{y}\right) - 2\arctan\left(\frac{x-3}{y}\right)\right] \\ &= \left[2\arctan\left(\frac{x+3}{y}\right) + \arctan\left(\frac{x-5}{y}\right) + \arctan\left(\frac{x+5}{y}\right)\right]. \end{aligned}$$

7. We will solve the following problem for the left quarter plane:

$$\nabla^2 u = 0 \text{ for } x < 0, y > 0,$$
$$u(0,y) = 0,\ u(x,0) = f(x) \text{ for } x < 0,\ y > 0.$$

Begin with a Dirichlet problem for the upper half plane, with $u(0,y)=0$ and $u(x,0)=g(x)$. We know how to solve this problem:

$$u_{hp}(x,y)=\frac{y}{\pi}\int_{-\infty}^{\infty}\frac{g(\xi)}{y^2+(\xi-x)^2}\,d\xi.$$

Now define

$$g(x)=\begin{cases}\text{any} & \text{for } x>0,\\ f(x) & \text{for } x<0.\end{cases}$$

We will fill in the "any" later. Write

$$u_{hp}(x,y)=\frac{y}{\pi}\left[\int_{-\infty}^{0}\frac{g(\xi)}{y^2+(\xi-x)^2}\,d\xi+\int_{0}^{\infty}\frac{g(\xi)}{y^2+(\xi-x)^2)}\,d\xi\right].$$

Put $w=-\xi$ in the second integral to obtain

$$\begin{aligned}u_{hp}(x,y)&=\frac{y}{\pi}\left[\int_{-\infty}^{0}\frac{g(\xi)}{y^2+(\xi-x)^2}\,d\xi+\int_{0}^{-\infty}\frac{g(-w)}{y^2+(w+x)^2}(-1)\,dw\right]\\&=\frac{y}{\pi}\left[\int_{-\infty}^{0}\frac{g(\xi)}{y^2+(\xi-x)^2}\,d\xi+\int_{-\infty}^{0}\frac{g(-\xi)}{y^2+(\xi+x)^2}\,d\xi\right].\end{aligned}$$

In the last integral, we replaced w with ξ as the "dummy" variable of integration to have the same variable in both integrals, making the notation more transparent. This replacement is independent of the change of variables in the preceding line. The last expression suggests that we should complete the definition of g by putting

$$g(x)=\begin{cases}-f(-x) & \text{for } x>0,\\ f(x) & \text{for } x<0.\end{cases}$$

Now

$$\begin{aligned}u_{hp}(x,y)&=\frac{y}{\pi}\left[\int_{-\infty}^{0}\frac{f(\xi)}{y^2+(\xi-x)^2}\,d\xi+\int_{-\infty}^{0}\frac{-f(\xi)}{y^2+(\xi+x)^2}\,d\xi\right]\\&=\frac{y}{\pi}\int_{-\infty}^{0}\left[\frac{1}{y^2+(\xi-x)^2}-\frac{1}{y^2+(\xi+x)^2}\right]f(\xi)\,d\xi.\end{aligned}$$

This satisfies Laplace's equation in the upper half plane, hence also in the left quarter plane, and also satisfies $u(x,0)=f(x)$ for $x<0$ and $u(0,y)=0$ for $y>0$.

8. One approach here is to adapt the solution for the lower half plane to the solution for the quarter plane $x<0, y<0$. First obtain the solution

$$u_{lhp}=-\frac{y}{\pi}\int_{-\infty}^{\infty}\frac{g(\xi)}{y^2+(\xi-x)^2}\,d\xi$$

for the lower half plane. Then obtain the solution for this quarter plane as

$$u(x,y) = -\frac{y}{\pi}\int_{-\infty}^{0}\left[\frac{1}{y^2+(\xi-x)^2} - \frac{1}{y^2+(\xi+x)^2}\right] f(\xi)\,d\xi.$$

It is easy to observe from this expression that $u(0,y) = 0$.

10. The solution is

$$\begin{aligned} u(x,y) &= \frac{y}{\pi}\int_{-\infty}^{0}\left[\frac{f(\xi)}{y^2+(\xi-x)^2} - \frac{f(\xi)}{y^2+(\xi+x)^2}\right] d\xi \\ &= \frac{y}{\pi}\int_{-5}^{0}\left[\frac{\xi}{y^2+(\xi-x)^2} - \frac{\xi}{y^2+(\xi+x)^2}\right] d\xi \\ &= -\frac{1}{2\pi}\left[y\ln(y^2+10x+x^2+25) - 2x\arctan\left(\frac{x+5}{y}\right)\right. \\ &\left. -y\ln(y^2-10x+x^2+25) + 2x\arctan\left(\frac{x-5}{y}\right)\right]. \end{aligned}$$

12. Separation of variables leads to the solution

$$u(x,y) = \alpha_0 + \int_0^\infty a_\omega \cos(\omega x) e^{-\omega y}\,d\omega,$$

where

$$a_\omega = -\frac{2}{\pi\omega}\int_0^\infty f(\xi)\cos(\omega\xi)\,d\xi$$

and α_0 is an arbitrary constant.

14. Imitate the argument used to solve the Neumann problem for the upper half plane to obtain the solution

$$u(x,y) = -\frac{1}{2\pi}\int_{-\infty}^{\infty} \ln(x^2+(\eta-y)^2) g(\eta)\,d\eta + c,$$

in which c is an arbitrary constant.

16. It is a routine differentiation to verify that $u_n(x,y)$ is a solution for each positive integer n. Suppose now that n is given. Choose $y = \pi/2n$, so $\sin(ny) = 1$. We get

$$u_n(x,\pi/2n) = \frac{1}{n^2}\sinh(nx) = \frac{1}{n^2}\left(e^{nx} - e^{-nx}\right)$$

which can be made as large as we like by choosing x sufficiently large. There are infinitely many other choices of x and y that can also be used.

Chapter 6

Solutions Using Eigenfunction Expansions

6.1 A Theory of Eigenfunction Expansions

2. An examination of cases leads to the eigenvalues and eigenfunctions

$$\lambda_n = \left(\frac{(2n-1)\pi}{4}\right)^2, \; y_n(x) = \sin\left(\frac{(2n-1)\pi x}{4}\right)$$

for $n = 1, 2, \ldots$. An eigenfunction expansion of $f(x) = 2x$ will have the form

$$\sum_{n=1}^{\infty} c_n y_n(x),$$

where

$$\begin{aligned} c_n &= \frac{\int_0^2 2\xi \sin(((2n-1)\pi\xi)/4)\, d\xi}{\int_0^2 \sin^2(((2n-1)\pi\xi)/4)\, d\xi} \\ &= \frac{32(-1)^{n+1}}{\pi^2(2n-1)^2}. \end{aligned}$$

This series converges to $f(x)$ for $0 < x < 2$. Figure 6.1 shows graphs of $f(x) = 2x$ and the sum of the first five terms of this eigenfunction expansion.

4. The eigenvalues and eigenfunctions are

$$\lambda_n = \left(\frac{n\pi}{3}\right)^2, y_n(x) = \cos\left(\frac{n\pi}{3}x\right)$$

Solutions Manual to Accompany Beginning Partial Differential Equations, Third Edition. Peter V. O'Neil.

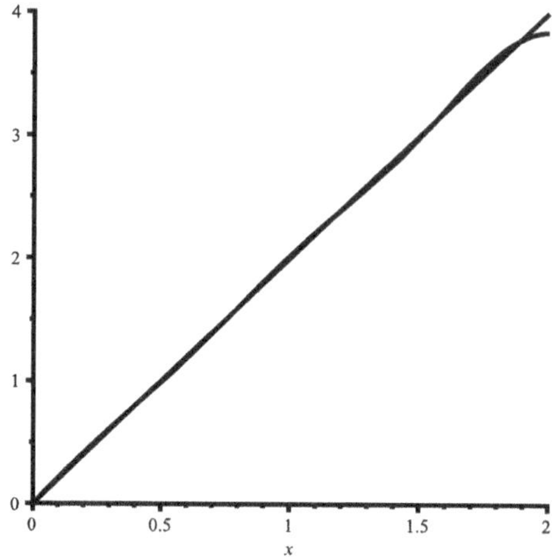

Figure 6.1: $f(x)$ and the 5th partial sum of the eigenfunction expansion in Problem 2.

for $n = 1, 2, \ldots$. An eigenfunction expansion of $f(x) = e^{-x}$ on $[0, 3]$ has the form

$$c_0 + \sum_{n=1}^{\infty} c_n \cos(n\pi x/3),$$

where

$$c_0 = \frac{\int_0^3 e^{-x}\,dx}{\int_0^3 dx} = \frac{1}{3}\left(e^{-3} - 1\right)$$

and, for $n = 0, 1, 2, \ldots$,

$$c_n = \frac{\int_0^3 e^{-x}\cos(n\pi x/3)\,dx}{\int_0^3 \cos^2(n\pi x/3)\,dx}$$
$$= \frac{2}{9 - n^2\pi^2}(1 - e^{-3}(-1)^n).$$

The series converges to e^{-x} for $0 < x < 3$. Figure 6.2 shows the function and the 10th partial sum of this expansion.

6. The differential equation is not in Sturm–Liouville form. Because the coefficient of y' is 1, multiply the differential equation by

$$e^{\int dx} = e^x$$

to obtain

$$e^x y'' + e^x y' + (1 + 4\lambda)e^x y = 0,$$

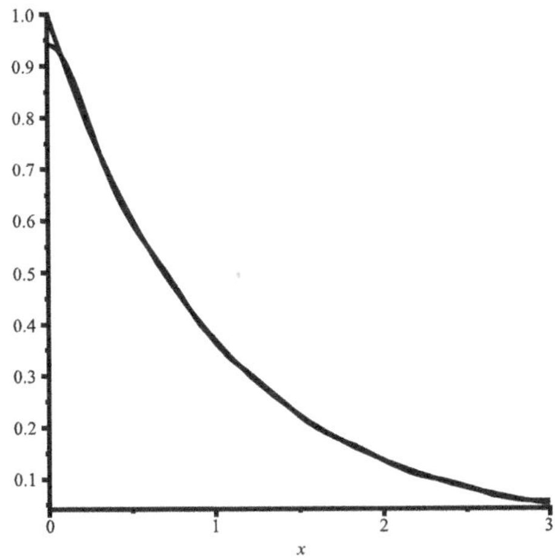

Figure 6.2: $f(x)$ and the 10th partial sum of the eigenfunction expansion in Problem 4.

or

$$(e^x y')' + (1 + 4\lambda)e^x y = 0.$$

The weight function in the orthogonality of the eigenfunctions for this problem is $p(x) = 4e^x$, the coefficient of λ. Now find the eigenvalues and eigenfunctions. For this, use the original form of the differential equation and attempt a solution $y = e^{rx}$. This requires that r be chosen as

$$r = \frac{-1 \pm \sqrt{1 - 4(1 + 4\lambda)}}{2} = \frac{-1 \pm \sqrt{-3 - 16\lambda}}{2}.$$

Take cases on $-3 - 16\lambda$. If $-3 - 16\lambda = 0$, then

$$y = ae^{-x/2} + bxe^{-x/2}.$$

Now $y(0) = y(1) = 0$ forces $a = b = 0$, so this case yields no eigenfunction.
If $-3 - 16\lambda > 0$, say $-3 - 16\lambda = \alpha^2$, then

$$y = ae^{(-1+\alpha)x/2} + be^{(-1-\alpha)x/2}.$$

Now check that $y(0) = y(1) = 0$ yields $a = b = 0$, so this case also yields no eigenfunction.
Finally, suppose $-3 - 16\lambda < 0$, say $-3 - 16\lambda = -\alpha^2$. Now

$$y = ae^{-x/2}\cos(\alpha x/2) + be^{-x/2}\sin(\alpha x/2).$$

Then $y(0) = a = 0$, so $y(x) = be^{-x/2}\sin(\alpha x/2)$. Now

$$y(1) = be^{-1/2}\sin(\alpha/2) = 0.$$

This equation can be satisfied with nonzero b by choosing $\alpha/2 = n\pi$. Then $\alpha = 2n\pi$, with $n = 1, 2, \ldots$. This gives us eigenvalues and eigenfunctions

$$\lambda_n = \frac{-3 + 4n^2\pi^2}{16}, \; y_n(x) = e^{-x/2}\sin(n\pi x).$$

To expand $f(x) = x$ in a series of these eigenfunctions, the weight function is $p(x) = 4e^x$. The coefficients are

$$\begin{aligned} c_n &= \frac{\int_0^1 4e^x x e^{-x/2}\sin(n\pi x)\,dx}{\int_0^1 4e^x \left(e^{-x/2}\sin(n\pi x)\right)^2\,dx} \\ &= \frac{\int_0^1 x e^{x/2}\sin(n\pi x)\,dx}{\int_0^1 \sin^2(n\pi x)\,dx} \\ &= -\frac{4}{(4n^2\pi^2+1)^2}\left(8n\pi + 8e^{1/2}n^3\pi^3(-1)^n - 6e^{1/2}n\pi(-1)^n\right). \end{aligned}$$

The eigenfunction expansion

$$\sum_{n=1}^{\infty} c_n e^{-x/2}\sin(n\pi x)$$

converges to x on $0 < x < 1$. Figure 6.3 shows $f(x) = x$ and the fiftieth partial sum of this expansion. Greater accuracy could be achieved by taking more terms in the partial sum.

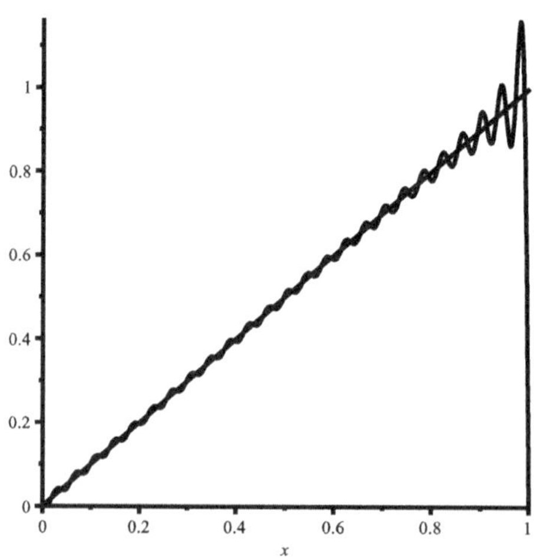

Figure 6.3: $f(x)$ and the 50th partial sum of the eigenfunction expansion in Problem 6.

7.

$$\int_a^b \int_a^b [f(x)g(y) - g(x)f(y)]^2 \, dx \, dy$$
$$= \int_a^b \int_a^b \left[(f(x))^2(g(y))^2 + (g(x))^2(f(y))^2 - 2f(x)g(x)f(y)g(y)\right] dx \, dy$$
$$= 2\int_a^b (f(x))^2 \, dx \int_a^b (g(y))^2 \, dy - 2\int_a^b f(x)g(x) \, dx \int_a^b f(y)g(y) \, dy$$
$$= 2 \parallel f \parallel^2 \parallel g \parallel^2 - 2(f \cdot g)^2.$$

Therefore

$$0 \leq \parallel f \parallel^2 \parallel g \parallel^2 - (f \cdot g)^2.$$

Then

$$(f \cdot g)^2 \leq \parallel f \parallel^2 \parallel g \parallel^2,$$

so

$$(f \cdot g) \leq \parallel f \parallel \parallel g \parallel .$$

6.2 Bessel Functions

2. Follow the template of equation 6.11. We need

$$2a - 1 = 7, \; b^2c^2 = 16, \; 2c - 2 = 2, \; \text{and } a^2 - \nu^2c^2 = 4.$$

Then $a = 4$, $c = 2$, $a = 4$ and $\nu = \sqrt{3}$. The solution is

$$y = c_1 x^4 J_{\sqrt{3}}(2x^2) + c_2 x^4 J_{-\sqrt{3}}(2x^2).$$

4. Using equation 6.11, obtain the solution

$$y = c_1 x^5 J_{\sqrt{5}/2}(4x^6) + c_2 x^5 J_{-\sqrt{5}/2}(4x2).$$

6. Let $u = x^{3/2}$ and $Y(u) = y(x(u))$. It is a straightforward exercise in chain rule differentiation to transform the differential equation for $y(x)$ to:

$$4u^{4/3}\left[\frac{9}{4}u^{2/3}Y'' + \frac{3}{4}u^{-1/3}Y'\right] + 4u^{2/3}\left[\frac{3}{2}u^{1/3}Y'\right] + (9u^2 - 16) = 0.$$

This simplifies to

$$u^2Y'' + uY' + (u^2 - 1)Y = 0.$$

This has general solution

$$Y(u) = c_1 J_2(u) + c_2 Y_2(u).$$

In terms of $y(x)$, this transforms to

$$y(x) = c_1 J_2(x^{3/2}) + c_2 Y_2(x^{3/2}).$$

8. With $y = x^{2/3}u$, the differential equation transforms to

$$36x^2\left[x^{2/3}u'' + \frac{4}{3}x^{-1/3}u' - \frac{2}{9}x^{-4/3}u\right]$$
$$- 12x\left[x^{2/3}u' + \frac{2}{3}x^{-1/3}\right] + (36x^2+7)x^{2/3}u = 0.$$

This simplifies to

$$x^2u'' + xu' + (x^2 - 1/4)u = 0,$$

which has the general solution

$$u(x) = c_1J_{1/2}(x) + c_2Y_{1/2}(x).$$

Then

$$y(x) = c_1x^{2/3}J_{1/2}(x) + c_2x^{2/3}Y_{1/2}(x).$$

9. Let α be a positive zero of $J_0(x)$. Then $J_0(\alpha) = 0$. Recall that $J_0'(x) = -J_1(x)$. Then

$$\int_1^\alpha J_0(s)\,ds = -J_0(x)]_0^\alpha = J_0(0) - J_0(\alpha) = 1$$

because $J_0(0) = 1$. Now make the change of variable $s = \alpha x$ in the integral to obtain

$$\alpha\int_0^1 J_1(\alpha x)\,dx = 1$$

and this is equivalent to what we wanted to show.

10. Let $u(x) = J_0(ax)$. Then

$$u' = aJ_0'(ax) \text{ and } u'' = a^2J_0''(ax).$$

Then

$$\begin{aligned} xu'' + u' + a^2xu &= a^2xJ_0''(ax) + aJ_0'(ax) + a^2J_0(ax) \\ &= a[axJ_0''(ax) + J_0'(ax) + axJ_0(ax)] = 0, \end{aligned}$$

in which we have used Bessel's equation of order $\nu = 0$. Similarly,

$$xv'' + v' + b^2xv = 0.$$

(b) Write

$$\begin{aligned} &v(xu'' + u' + a^2xu) - u(xv'' + v' + b^2xv) \\ =\ &xvu'' - xuv'' + vu' - uv' + (a^2-b^2)xuv = 0. \end{aligned}$$

This equation can be written as

$$(b^2 - a^2)(xuv) = [x(u'v - v'u)]'.$$

(c) Integrate both sides of the equation derived in part (b) to obtain

$$(b^2 - a^2) \int xuv\, dx = x(vu' - uv').$$

With $u = J_0(ax)$ and $v = J_0(bx)$, this yields Lommel's integral

$$(b^2 - a^2) \int x J_0(ax) J_0(bx)\, dx = x\left(a J_0'(ax) J_0(bx) - b J_0'(bx) J_0(ax)\right).$$

11. Begin with equation 6.12:

$$(x^n J_n(x))' = x^n J_{n-1}(x).$$

Upon integrating, we obtain

$$\int x^n J_{n-1}(x)\, dx = x^n J_n(x).$$

Similarly, by equation 6.13,

$$(x^{-n} J_n(x))' = -x^{-n} J_{n+1}(x),$$

so

$$\int x^{-n} J_{n+1}(x)\, dx = -x^{-n} J_n(x).$$

12. The Fourier–Bessel expansion has the form

$$\sum_{n=1}^{\infty} c_n J_0(j_n x),$$

where j_n is the nth positive zero of $J_0(x)$ and

$$c_n = \frac{2}{(J_1(j_n))^2} \int_0^1 -2\xi J_0(j_n \xi)\, d\xi.$$

The series converges to -2 for $0 < x < 1$. Figure 6.4 shows a graph of $f(x) = -2$ and the 50th partial sum of this expansion.

14. The Fourier–Bessel expansion has the form

$$\sum_{n=1}^{\infty} c_n J_0(j_n x),$$

where, as usual, j_n is the nth positive zero of $J_0(x)$ and

$$c_n = \frac{2}{(J_1(j_n))^2} \int_0^1 x(x^2 - x) J_0(j_n x)\, dx.$$

In this example, the 10th partial sum appears to approximate the function well on $[0, 1]$ (Figure 6.5).

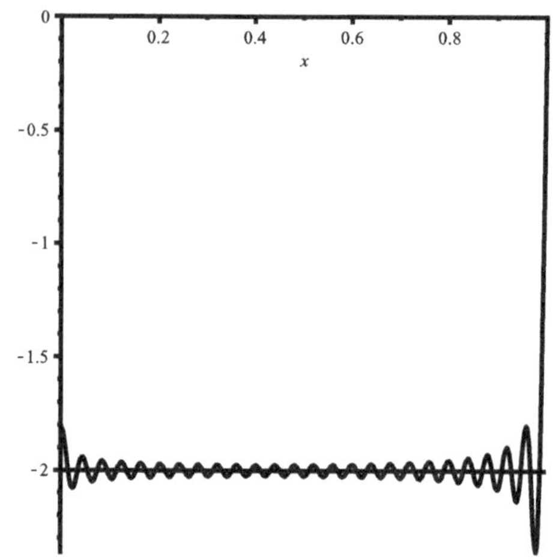

Figure 6.4: 50th partial sum of the Fourier–Bessel expansion of $f(x) = -2$, Problem 12.

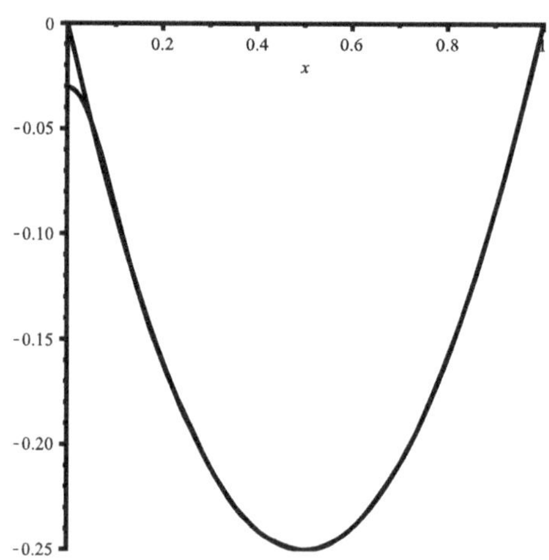

Figure 6.5: Tenth partial sum of the Fourier–Bessel expansion of $f(x) = x^2 - x$, Problem 14.

16. Begin with the expansion

$$\cos(x\sin(\theta)) = \sum_{n=0}^{\infty} \frac{(-1)^n}{(2n)!} x^{2n} \sin^{2n}(\theta).$$

Now integrate term by term:

$$\int_0^\pi \cos(x\sin(\theta))\,d\theta = \sum_{n=0}^{\infty} \frac{(-1)^n}{(2n)!} x^{2n} \int_0^\pi \sin^{2n}(\theta)\,d\theta.$$

Now use known integral

$$\int_0^\pi \sin^{2n}(\theta)\,d\theta = \frac{(2n)!\pi}{2^{2n}(n!)^2},$$

which can be found in standard tables or evaluated by, for example, using residues in complex function theory. We obtain

to write

$$\frac{1}{\pi}\int_0^\pi \cos(x\sin(\theta))\,d\theta = \sum_{n=0}^{\infty} \frac{1}{2^{2n}(n!)^2} x^{2n} = J_0(x).$$

6.3 Applications of Bessel Functions

6.3.1 Temperature Distribution in a Solid Cylinder

2. The solution is

$$U(r,t) = \sum_{n=1}^{\infty} c_n J_0\left(\frac{j_n r}{3}\right) e^{-2j_n^2 t/9},$$

where j_n is the nth positive zero of $J_0(x)$ and

$$c_n = \frac{2}{(J_1(j_n))^2} \int_0^1 r\sin(3\pi r) J_0(j_n r)\,dr.$$

Figure 6.6 shows graphs of the solution at a selection of times.

4. The solution is

$$U(r,t) = \sum_{n=1}^{\infty} c_n J_0(j_n r) e^{-2j_n^2 t},$$

where

$$c_n = \frac{2}{(J_1(j_n))^2} \int_0^1 r(1+\cos(\pi r)) J_0(j_n r) e^{-2j_n^2 t}.$$

Figure 6.7 shows graphs of $U(r,t)$ for $t = 0, 1/20, 1/10, 1/2$.

6.3.2 Vibrations of a Circular Drum

2. The solution is

$$z(r,t) = \sum_{n=1}^{\infty} c_n J_0\left(\frac{j_n}{2} r\right) \cos\left(\frac{3j_n}{2} t\right),$$

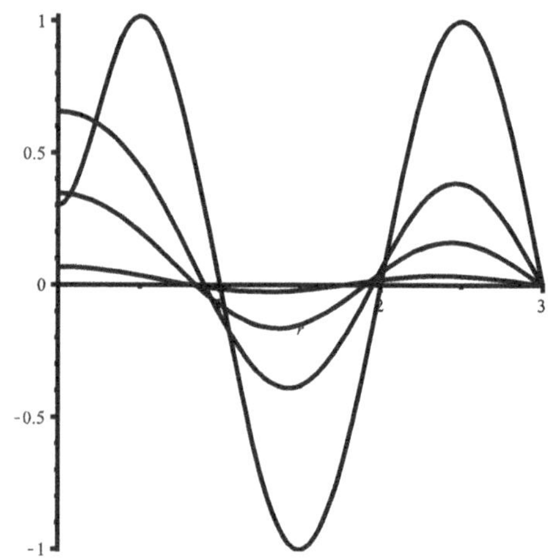

Figure 6.6: Graphs of $U(r,t)$ for $t = 0, 1/5, 1/10, 1/20$ in Problem 2.

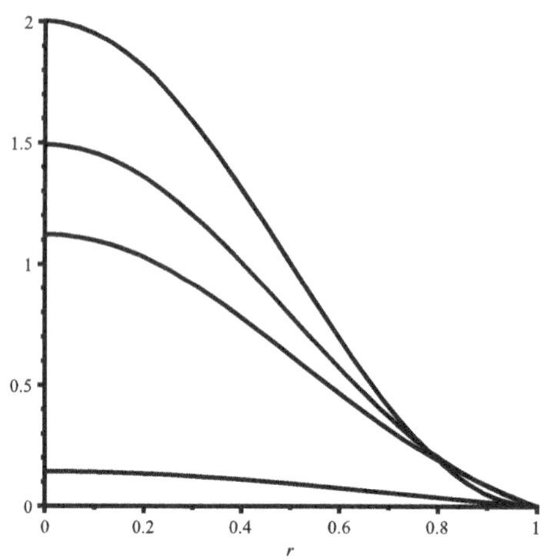

Figure 6.7: Graphs of $U(r,t)$ for $t = 0, 1/20, 1/10, 1/2$ in Problem 4.

where

$$c_n = \frac{2}{J_1^2(j_n)} \int_0^1 r \sin(2\pi r) J_0(j_n r)\, dr.$$

Figure 6.8 shows graphs of $z(r,t)$ at $t = 0, 1/20, 1/10, 1/2$.

4. The solution is

$$z(r,t) = \sum_{n=1}^{\infty} c_n J_0\left(\frac{j_n}{2} r\right) \cos(j_n t),$$

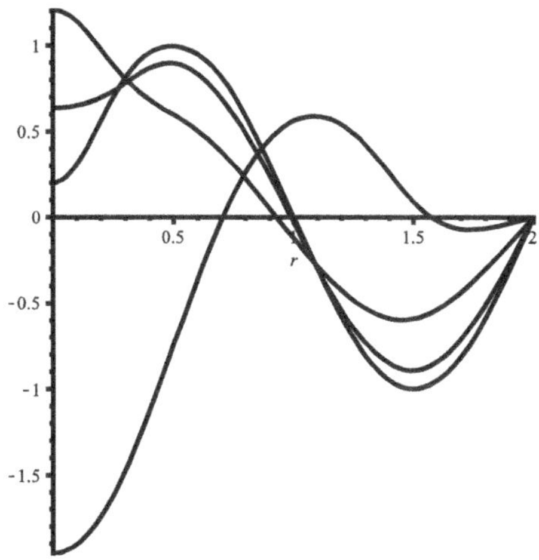

Figure 6.8: Graphs of $z(r,t)$ for $t = 0, 1/20, 1/10, 1/2$ in Problem 2.

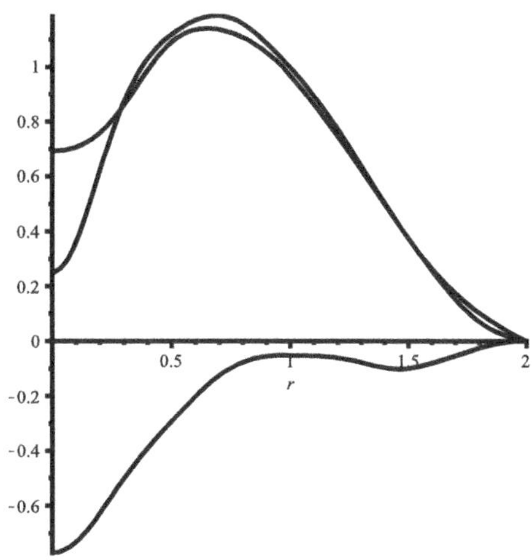

Figure 6.9: Graphs of $z(r,t)$ for $t = 0, 1/20, 1/10, 1/2$ in Problem 4.

with

$$c_n = \frac{2}{J_1^2(j_n)} \int_0^1 2r^2(2-2r)^2 J_0(j_n r)\, dr.$$

Figure 6.9 shows graphs of $z(r,t)$ at times $t = 0, 1/20, 1/10, 1/2$.

6.4 Legendre Polynomials and Applications

1. Write
$$\frac{1}{\sqrt{1-w}} = 1 + \frac{1}{2}w + \frac{3}{8}w^2 + \frac{15}{48}w^3 + \cdots .$$
Put $w = 2xt - t^2$ to obtain, after some algebra,
$$\begin{aligned}\frac{1}{\sqrt{1-w}} =& 1 + xt - \frac{1}{2}t^2 + \frac{3}{2}x^2t^2 - \frac{3}{2}xt^3 + \frac{3}{8}t^4 \\ &+ \frac{5}{2}x^3t^3 - \frac{15}{4}x^2t^4 + \frac{15}{8}xt^5 - \frac{5}{16}t^6 + \cdots .\end{aligned}$$
Collect the coefficients of t, t^2, t^3 to write
$$\begin{aligned}\frac{1}{\sqrt{1-w}} &= 1 + xt \\ &+ \left(-\frac{1}{2} + \frac{3}{2}x^2\right)t^2 + \left(-\frac{3}{2}x + \frac{5}{2}x^3\right)t^3 + \cdots .\end{aligned}$$
From this abbreviated list of terms, we can read the first four Legendre polynomials, with $P_n(x)$ the coefficient of t^n. This gives us
$$P_0(x) = 1, P_1(x) = x, P_2(x) = -\frac{1}{2} + \frac{3}{2}x^2, P_3(x) = -\frac{3}{2}x + \frac{5}{x}x^3.$$

2. We know that, for $-1 < t < 1$,
$$\frac{1}{\sqrt{1 - 2xt + t^2}} = \sum_{n=0}^{\infty} P_n(x)t^n.$$
Put $x = t = 1/2$ in this to obtain
$$\frac{1}{\sqrt{3/4}} = \sum_{n=0}^{\infty} P_n\left(\frac{1}{2}\right)\frac{1}{2^n}.$$
Then
$$\sum_{n=0}^{\infty} \frac{1}{2^n} P_n\left(\frac{1}{2}\right) = \frac{2}{\sqrt{3}}.$$

3. From Figure 6.15 of the text and the law of cosines,
$$R^2 = r^2 + d^2 - 3rd\cos(\theta),$$
so
$$\frac{R^2}{d^2} = 1 - 2\frac{r}{d}\cos(\theta) + \frac{r^2}{d^2}.$$
Then
$$\varphi(x, y, z) = \frac{1}{R} = \frac{1}{d\sqrt{1 - 2\left(\frac{r}{d}\right)\cos(\theta) + \left(\frac{r}{d}\right)^2}}.$$

(b) Suppose $r/d < 1$. By comparing the result in part a with the generating function for the Legendre polynomials, with $x = \cos(\theta)$ and $t = r/d$, we have

$$\varphi(r) = \frac{1}{d}\sum_{n=0}^{\infty} \frac{1}{d^{n+1}} P_n(\cos(\theta)) r^n.$$

(c) Now suppose that $r/d > 1$. In this case write

$$\frac{R^2}{r^2} = 1 - 2\frac{d}{r}\cos(\theta) + \frac{d^2}{r^2}.$$

Then

$$\frac{r}{R} = \frac{1}{\sqrt{1 - 2\frac{d}{r}\cos(\theta) + \frac{d^2}{r^2}}}.$$

Again comparing with the generating function, we have

$$\varphi(r) = \frac{1}{r}\sum_{n=0}^{\infty} P_n(\cos(\theta)) \left(\frac{d}{r}\right)^n.$$

This can be written as

$$\varphi(r) = \frac{1}{r}\sum_{n=0}^{\infty} d^n P_n(\cos(\theta)) r^{-n}.$$

4. The fact that $P_{2n+1}(0) = 0$ is obvious because, the way the Legendre polynomials are defined, each odd-order polynomial has a factor of x.

   For even order Legendre polynomials, we can use the recurrence relation to carry out an inductive argument. First replace each n with $2n - 1$ in the recurrence relation to write

$$2nP_{2n}(x) - (4n+1)xP_{2n-1}(x) + (2n-1)P_{2n-2}(x) = 0.$$

   Then

$$P_{2n}(x) = -\frac{2n-1}{2n} P_{2n-2}(x).$$

   If the expression is true for $P_{2n-2}(0)$, then

$$\begin{aligned} P_{2n}(0) &= -\frac{2n-1}{2n}\left[\frac{(2(n-1))!}{2^{2(n-1)}((n-1)!)^2}\right](-1)^{n-1} \\ &= \frac{(2n-1)!}{2^{2n-1}n!(n-1)!}(-1)^n \\ &= \frac{(2n)!}{2^{2n}(n!)^2}(-1)^n. \end{aligned}$$

6. For this expansion we need only the first four Legendre polynomials. Write

$$6x + x^2 - 4x^3 = c_0 P_0(x) + c_1 P_1(x) + c_2 P_2(x) + c_3 P_3(x).$$

The coefficients on the right can be obtained by straightforward algebraic manipulation, comparing coefficients of like powers of x, or by computing them as the Fourier–Legendre coefficients of the polynomial on the left. Either way we obtain

$$6x + x^2 - 4x^3 = \frac{1}{3}P_0(x) + \frac{18}{5}P_1(x) + \frac{2}{3}P_2(x) - \frac{8}{5}P_3(x).$$

8. With $f(x) = |x|$, the 11th partial sum of the Fourier–Legendre expansion is

$$S_{11}(x) = \sum_{n=0}^{10} c_n P_n(x),$$

where

$$c_n = \frac{2n+1}{2}\int_{-1}^{1} |\xi| P_n(\xi)\, d\xi.$$

Figure 6.10 compares $f(x)$ with $S_{11}(x)$ on $[-1, 1]$.

10. The expansion is

$$\sum_{n=0}^{\infty} c_n P_n(x),$$

where

$$c_n = \frac{2n+1}{2}\int_{-1}^{1} x\sin(x) P_n(x)\, dx.$$

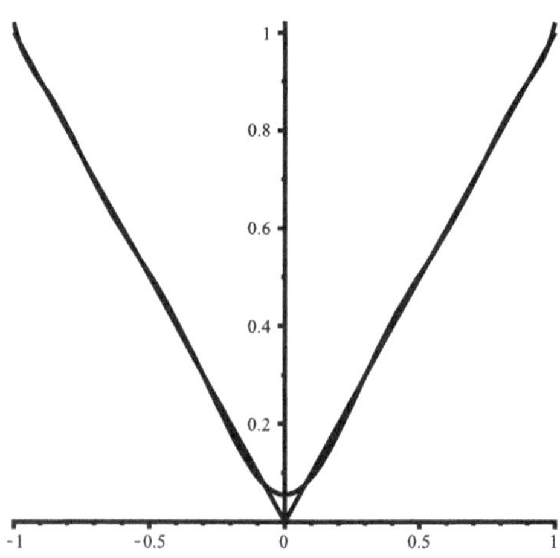

Figure 6.10: Eleventh partial sum of the Fourier–Legendre expansion of $|x|$.

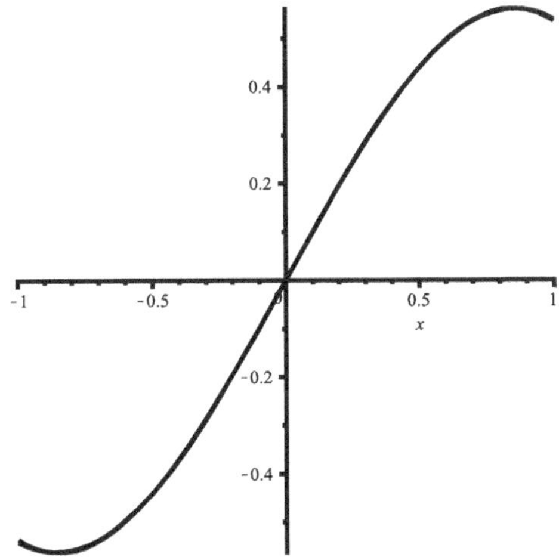

Figure 6.11: Eleventh partial sum of the Fourier–Legendre expansion of $x\cos(x)$.

12. The expansion is

$$\sum_{n=1}^{\infty} c_n P_n(x),$$

where

$$c_n = \frac{2n+1}{2}\int_{-1}^{1} x\cos(x)P_n(x)\,dx.$$

Figure 6.11 shows a graph of the function and the 11th partial sum of this expansion.

14. The solution is

$$u(\rho,\varphi) = \sum_{n=0}^{\infty} c_n \rho^n P_n(\cos(\varphi)),$$

in which, with $f(\varphi) = \varphi$,

$$c_n = \frac{2n+1}{2}\int_{-1}^{1} \arccos(\xi)P_n(\xi)\,d\xi.$$

Figure 6.12 is a rectangular plot of $u(\rho,\varphi)$.

16. The solution is

$$u(\rho,\varphi) = \sum_{n=0}^{\infty} c_n \left(\frac{\rho}{2}\right)^n P_n(\cos(\varphi)),$$

where

$$c_n = \frac{2n+1}{2}\int_{-1}^{1} (\arccos^2(\xi) - \arccos(\xi))P_n(\cos(\xi))\,d\xi.$$

Figure 6.13 is a rectangular plot of $u(\rho,\varphi)$.

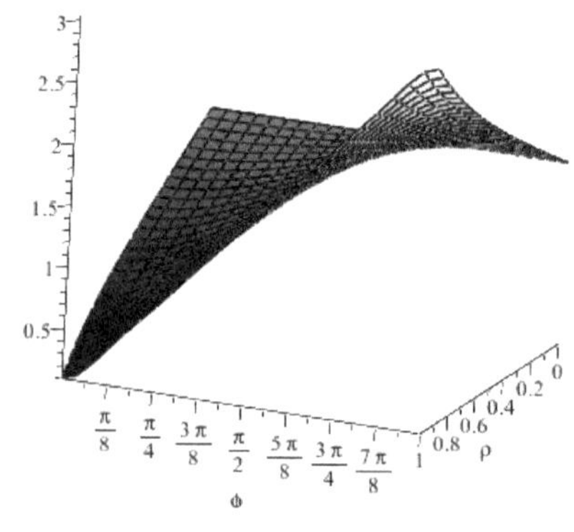

Figure 6.12: Plot of $u(\rho, \varphi)$ (in rectangular format) in Problem 14.

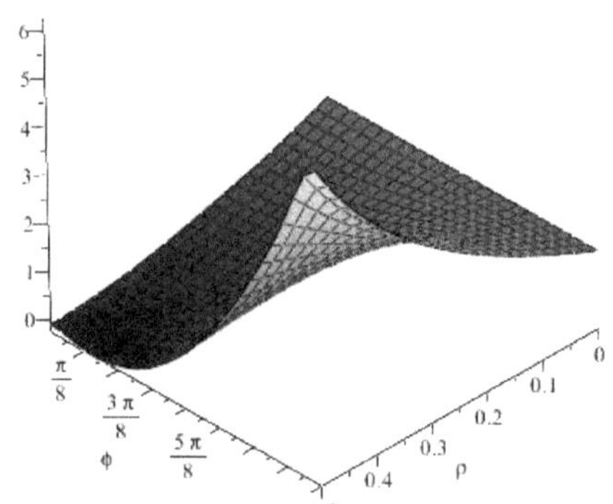

Figure 6.13: Rectangular plot of $u(\rho, \varphi)$ in Problem 16.

18. The solution is

$$u(\rho, \varphi) = \sum_{n=0}^{\infty} c_n \left(\frac{\rho}{3}\right)^n P_n(\cos(\varphi)),$$

where

$$c_n = \frac{2n+1}{2} \int_{-1}^{1} \sin^2(\arccos(\xi)) P_n(\cos(\xi))\, d\xi.$$

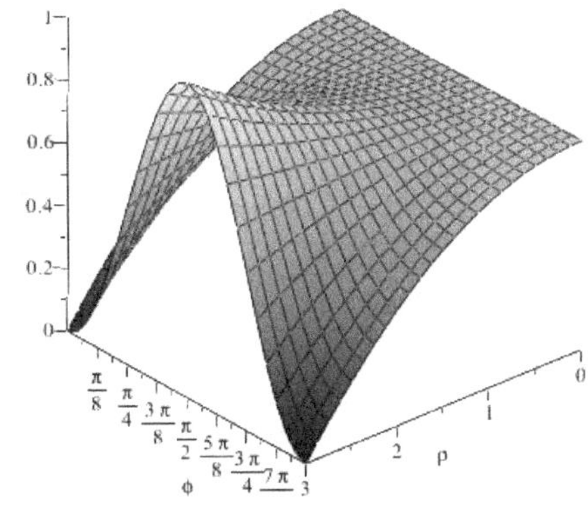

Figure 6.14: Rectangular plot of $u(\rho, \varphi)$ in Problem 18.

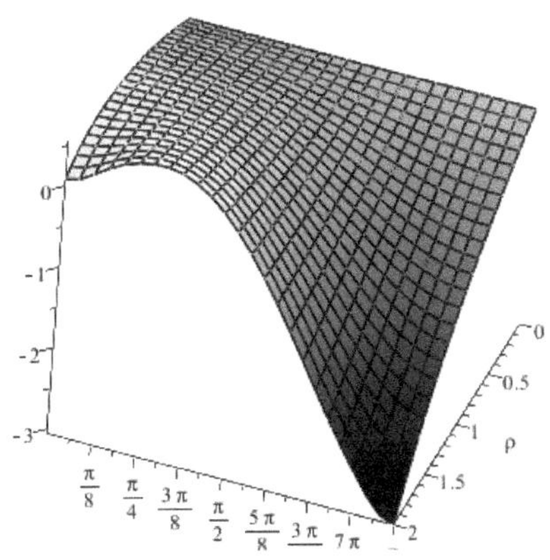

Figure 6.15: Rectangular plot of $u(\rho, \varphi)$ in Problem 20.

Figure 6.14 is a rectangular plot of this solution.

20. The solution is

$$u(\rho, \varphi) = \sum_{n=0}^{\infty} c_n \left(\frac{\rho}{2}\right)^n P_n(\cos(\varphi)),$$

where

$$c_n = \frac{2n+1}{2} \int_{-1}^{1} (\xi \arccos(\xi) P_n(\xi))\, d\xi.$$

Figure 6.15 is a rectangular plot of $u(\rho, \varphi)$.

Chapter 7

Integral Transform Methods of Solution

7.1 The Fourier Transform

Problems 2–10 were done using MAPLE.

2.

$$\widehat{f}(\omega) = \frac{2}{\omega^2 + 1}.$$

4.

$$\widehat{f}(\omega) = \frac{2}{\omega^2}\left(a\omega\sin(a\omega) - 2\sin^2(a\omega/2)\right).$$

6.

$$\widehat{f}(\omega) = -\frac{4ik}{\omega}\sin^2(a\omega/2).$$

8.

$$\begin{aligned}\widehat{f}(\omega) &= \frac{1}{2}\frac{1}{\omega^2 - 1}\left(4i\omega\sin(a)\cos(a)e^{-2ia\omega} + 2i\omega\sin(a)e^{ia\omega}\right)\\ &\quad + \frac{1}{2}\frac{1}{\omega^2 - 1}\left(-e^{ia(\omega-1)} - e^{ia(\omega+1)} + e^{-2ia(\omega-1)} + e^{-2ia(\omega+1)}\right).\end{aligned}$$

9.

$$\widehat{f}(\omega) = \frac{2k}{k^2 + \omega^2}.$$

10.

$$\widehat{f}(\omega) = \frac{2(2 + \omega^2)}{4 + \omega^4}.$$

Solutions Manual to Accompany Beginning Partial Differential Equations,
Third Edition. Peter V. O'Neil.

13. The general formula follows by a straightforward induction if it is first proved for $n = 1$. To do this, integrate by parts and use the fact that $f(t) \to 0$ as $t \to \infty$ and as $t \to -\infty$:

$$\begin{aligned}(F)[f'(t)](\omega) &= \int_{-\infty}^{\infty} f'(\xi)e^{-i\omega\xi}\,d\xi \\ &= e^{-i\omega\xi}f(\xi)\Big]_{-\infty}^{\infty} - \int_{-\infty}^{\infty} f(\xi)(-i\omega)e^{-i\omega\xi}\,d\xi \\ &= i\omega\int_{-\infty}^{\infty} f(\xi)e^{-i\omega\xi}\,d\xi \\ &= i\omega\widehat{f}(\omega).\end{aligned}$$

14. Suppose $p > 0$. Begin with

$$\mathcal{F}[pf(t)](\omega) = \int_{-\infty}^{\infty} pf(p\xi)e^{-i\omega\xi}\,d\xi.$$

Make the change of variable $p\xi = u$ to obtain

$$\begin{aligned}\mathcal{F}[pf(t)](\omega) &= \int_{-\infty}^{\infty} pf(\omega)e^{-i\omega u/p}\frac{1}{p}\,du \\ &= \int_{-\infty}^{\infty} f(u)e^{-i\omega u/p}\,du \\ &= \widehat{f}(\omega/p).\end{aligned}$$

15. Begin with

$$\mathcal{F}[f(t-t_0)](\omega) = \int_{-\infty}^{\infty} f(\xi-t_0)e^{-i\omega\xi}\,d\xi.$$

Let $u = \xi - t_0$ to obtain

$$\begin{aligned}\mathcal{F}[f(t-t_0)](\omega) &= \int_{-\infty}^{\infty} f(u)e^{-i\omega(u+t_0)}\,du \\ &= \int_{-\infty}^{\infty} f(u)e^{-i\omega t_0}e^{-i\omega u}\,du \\ &= e^{-i\omega t_0}\int_{-\infty}^{\infty} f(u)e^{-i\omega u}\,du \\ &= e^{-i\omega t_0}\widehat{f}(\omega).\end{aligned}$$

16.

$$\begin{aligned}\mathcal{F}[e^{ikt}f(t)](\omega) &= \int_{-\infty}^{\infty} e^{ik\xi}f(\xi)e^{-i\omega\xi}\,d\xi \\ &= \int_{-\infty}^{\infty} f(\xi)e^{-i\xi(\omega-k)}\,d\xi \\ &= \widehat{f}(\omega-k).\end{aligned}$$

17. This equation can be verified by considering cases on k and using the scaling property of the transform. First suppose that $k > 0$. Then $|k| = k$ and we have immediately that

$$\mathcal{F}(kf(kt)](\omega) = \widehat{f}(\omega/k),$$

so

$$\mathcal{F}[f(kt)](\omega) = \frac{1}{k}\widehat{f}(\omega/k) = \frac{1}{|k|}\widehat{f}(\omega/k).$$

Now suppose that $k < 0$, so $|k| = -k$. Replace k with $-k$ in the scaling property to write

$$\mathcal{F}[f(-kt)](\omega) = \frac{1}{-k}\widehat{f}(\omega/-k).$$

Then

$$\mathcal{F}[f(-kt)](\omega) = \frac{1}{-k}\widehat{f}(\omega/k) = \frac{1}{|k|}\widehat{f}(\omega/k).$$

18. To prove time reversal, begin with

$$\mathcal{F}[f(-t)](\omega) = \int_{-\infty}^{\infty} f(-\xi)e^{-i\omega\xi}\,d\xi.$$

Let $u = -\xi$ to continue

$$\begin{aligned}\mathcal{F}[f(-t)](\omega) &= \int_{\infty}^{-\infty} f(u)e^{i\omega u}(-1)\,du \\ &= \int_{-\infty}^{\infty} f(u)e^{-i(-\omega)u}\,du \\ &= \widehat{f}(-\omega).\end{aligned}$$

19. To prove the symmetry property, write

$$\begin{aligned}\mathcal{F}[\widehat{f}(t)](\omega) &= \int_{-\infty}^{\infty} \widehat{f}(\xi)e^{-i\omega\xi}\,d\xi \\ &= \int_{-\infty}^{\infty} \widehat{f}(\xi)e^{i(-\omega)\xi}\,d\xi \\ &= 2\pi f(-\omega)\end{aligned}$$

by the inversion formula (assuming continuity of f).

20. A straightforward integration (using the definition of convolution), yields the result

$$(f * g)(t) = 2e^{-1-t}.$$

21. It is necessary to consider separately the cases $k \neq 1$ and $k = 1$.

    If $k \neq 1$, compute the following integrations (with routine details omitted). For $t < 0$,

$$
\begin{aligned}
(f * g)(t) &= \int_{-\infty}^{t} f(t-\xi)g(\xi)\,d\xi + \int_{t}^{0} f(t-\xi)g(\xi)\,d\xi \\
&\quad + \int_{0}^{\infty} f(t-\xi)g(\xi)\,d\xi \\
&= \frac{1}{k+1}\left(e^{t} + e^{kt}\right) + \frac{1}{k-1}\left(1 - e^{(k-1)t}\right).
\end{aligned}
$$

    And, if $t > 0$,

$$
(f * g)(t) = \frac{1}{k+1}\left(e^{-t} + e^{-kt}\right) + \frac{1}{k-1}\left(1 - e^{(1-k)t}\right).
$$

    In the case that $k = 1$ we have $f = g$ and we obtain

$$
(f * f)(t) = \begin{cases} e^{t} = te^{t} & \text{if } t < 0, \\ e^{-t} + te^{-t} & \text{if } t > 0. \end{cases}
$$

22. Integrations yield

$$
(f * f)(t) = \begin{cases} 0 & \text{if } t \leq -2k, \\ -\frac{2}{3}k^3 - k^2 t + \frac{1}{6}t^3 & \text{if } -2k < t < 0. \\ -\frac{2}{3}k^3 + k^2 t - \frac{1}{6}t^3 & \text{if } 0 < t < 2k, \\ 0 & \text{if } t > 2k. \end{cases}
$$

23. First observe that

$$
\begin{aligned}
\int_{-\infty}^{\infty} (f * g)(t)\,dt &= \int_{-\infty}^{\infty} \left(\int_{-\infty}^{\infty} f(t-x)g(x)\,dx \right) dt \\
&= \int_{-\infty}^{\infty} \left(\int_{-\infty}^{\infty} f(t-x)\,dt \right) g(x)\,dx \\
&= \int_{-\infty}^{\infty} \left(\int_{-\infty}^{\infty} f(t)\,dt \right) g(x)\,dx \\
&= \int_{-\infty}^{\infty} f(x)\,dx \int_{-\infty}^{\infty} g(x)\,dx.
\end{aligned}
$$

    Next, we can also write

$$
\begin{aligned}
|(f * g)(t)| &= \left| \int_{-\infty}^{\infty} f(t-x)g(x)\,dx \right| \\
&\leq \int_{-\infty}^{\infty} |f(t-x)||g(x)|\,dx = (|f| * |g|)(t).
\end{aligned}
$$

Putting these two observations together, we have

$$\int_{-\infty}^{\infty} |(f * g)|(t)\,dt \le \int_{-\infty}^{\infty} (|f| * |g|)\,dt = \int_{-\infty}^{\infty} |f(t)|\,dt \int_{-\infty}^{\infty} |g(t)|\,dt.$$

Different conventions are used in defining Fourier sine and cosine transforms. Instead of having the factor of $2/\pi$ with the inversion formulas (as is done here), sometimes a factor of $\sqrt{2/\pi}$ is assigned to both the transforms and the inversion formulas. This means that an inverse sine or cosine transform computed by MAPLE must be multiplied by $\sqrt{\pi/2}$ to obtain the transform as defined here.

26.

$$\widehat{f}_S(\omega) = \frac{\omega}{1+\omega^2} \text{ and } \widehat{f}_C = \frac{1}{1+\omega^2}.$$

28.

$$\widehat{f}_S(\omega) = \frac{\omega - \omega\cos(a)\cos(a\omega) - \sin(a)\sin(a\omega)}{\omega^2 - 1}$$

and

$$\widehat{f}_C = \frac{\omega\cos(a)\sin(a\omega) - \sin(a)\cos(a\omega)}{\omega^2 - 1}.$$

30.

$$\widehat{f}_S(\omega) = \frac{\omega(\omega^2 + 1 - a^2)}{((\omega+a)^2+1)((\omega-a)^2+1)}$$

and

$$\widehat{f}_C(\omega) = \frac{1}{2(1+(\omega+a)^2)} + \frac{1}{2((\omega-a)^2+1)}.$$

7.2 Heat and Wave Equations

2. Because a boundary condition is given for $u_x(0,t)$ instead of $u(0,t)$, try the Fourier cosine transform in the x-variable. Using the operational formula, the transform of the wave equation is

$$\begin{aligned}\widehat{u}_C'(\omega,t) &= k\left(-\omega^2\widehat{u}_C(\omega,t) - u_x(0,t)\right)\\ &= \widehat{u}_C'(\omega,t) = -k\omega^2\widehat{u}_C(\omega,t).\end{aligned}$$

Here the cosine transform has passed through the derivative with respect to t, which is denoted by a prime. Then

$$\widehat{u}_C'(\omega,t) + k\omega^2\widehat{u}_C(\omega,t) = 0.$$

This has the general solution

$$\widehat{u}_C(\omega,t) = A_\omega e^{-\omega^2 kt} + B_\omega e^{\omega^2 kt}.$$

For a bounded solution, let $B_\omega = 0$ to obtain

$$\widehat{u}_C(\omega,t) = A_\omega e^{-\omega^2 kt}.$$

Apply the cosine transform to the initial condition to obtain

$$\widehat{u}_C(0,t) = A_\omega = \widehat{f}_C(\omega).$$

Then

$$\widehat{u}_C(\omega,t) = \widehat{f}(\omega)e^{-\omega^2 kt}.$$

This is the Fourier cosine transform of the solution. The solution is the inverse Fourier cosine transform of this expression:

$$\begin{aligned} u(x,t) &= \frac{2}{\pi}\int_0^\infty \widehat{f}(\omega)e^{-\omega^2 kt}\cos(\omega x)\,d\omega \\ &= \int_0^\infty \int_0^\infty f(\xi)\cos(\omega\xi)\cos(\omega x)e^{-\omega^2 kt}\,d\xi\,d\omega. \end{aligned}$$

4. Because of the form of the boundary condition, use the Fourier sine transform in x. With primes denoting differentiation with respect to time, use the operational formula for the transform of the u_{xx} term to obtain

$$\widehat{u}_S'(\omega,t) = -k\omega^2\widehat{u}_S(\omega,t) + \omega\widehat{u}_S(0,t) - t\widehat{u}_S(\omega,t).$$

Then

$$\widehat{u}_S' + k\omega^2\widehat{u}_S = -t\widehat{u}_S.$$

This can be written as

$$\widehat{u}_S' + (k\omega^2 + t)\widehat{u}_S = 0.$$

Think of this as a first-order linear ordinary differential equation for $\widehat{u}_S(\omega,t)$, with t the variable and ω carried along as a parameter. Multiply the equation by the integrating factor

$$e^{\int (k\omega^2+t)\,dt} = e^{k\omega^2 t + t^2/2}$$

to write the differential equation as

$$\left(e^{k\omega^2 t+t^2/2}\widehat{u}_S\right)' = 0.$$

This has solutions

$$e^{k\omega^2 t+t^2/2}\widehat{u}_S = C_\omega,$$

so

$$\widehat{u}_S(\omega,t) = C_\omega e^{-k\omega^2 t - t^2/2} = C_\omega e^{-t^2/2}e^{-k\omega^2 t}.$$

With $u(x,0) = xe^{-x}$, we have

$$\begin{aligned} C_\omega &= \widehat{u}_S(\omega,0) \\ &= \mathcal{F}_S\left[xe^{-x}\right](\omega,0) = \frac{2\omega}{(1+\omega^2)^2}. \end{aligned}$$

Then

$$\widehat{u}_S(\omega,t) = \frac{2\omega}{(1+\omega^2)} e^{-k\omega^2 t - t^2/2}.$$

The solution is

$$\begin{aligned} u(x,t) &= \mathcal{F}_S\left[\widehat{u}_S(\omega,t)\right](x,t) \\ &= \frac{2}{\pi} e^{-t^2/2} \int_0^\infty \frac{2\omega}{(1+\omega^2)} \sin(\omega x) e^{-k\omega^2 t}\, d\omega. \end{aligned}$$

6. Compute

$$\widehat{\varphi}(\omega) = \frac{4}{\omega^3}(\sin(\omega) - \omega\cos(\omega)).$$

The solution is

$$u(x,t) = \frac{1}{2\pi}\int_{-\infty}^{\infty} \widehat{\varphi}(\omega)\cos(\omega ct) e^{i\omega x}\, d\omega.$$

Since the solution must be real valued, we can use Euler's formula to replace $e^{i\omega x}$ with $\cos(\omega x) + i\sin(\omega x)$ and obtain

$$u(x,t) = \frac{1}{2\pi}\int_{-\infty}^{\infty} \frac{4}{\omega^3}(\sin(\omega) - \omega\cos(\omega))\cos(\omega x)\cos(\omega ct)\, d\omega.$$

8. Here

$$\widehat{\varphi}(\omega) = \frac{4}{\omega^2}\sin^2(2\omega).$$

The solution is

$$u(x,t) = \frac{1}{2\pi}\int_{-\infty}^{\infty} \frac{4}{\omega^2}\sin^2(2\omega)\cos(\omega ct)\cos(\omega x)\, d\omega.$$

10. With

$$\widehat{\varphi}(\omega) = \frac{4\sin(4\pi\omega)}{\omega(4-\omega^2)},$$

the solution is

$$u(x,t) = \frac{1}{2\pi}\int_{-\infty}^{\infty} \frac{4\sin(4\pi\omega)}{\omega(4-\omega^2)}\cos(\omega ct)\cos(\omega x)\, d\omega.$$

12. The Fourier transform of the initial velocity function is

$$\widehat{\psi}(\omega) = \frac{2i}{\omega^2}(5\omega\cos(5\omega) - \sin(5\omega)).$$

The solution is

$$u(x,t) = \frac{1}{2\pi}\int_{-\infty}^{\infty} \frac{1}{\omega c}\frac{2i}{\omega^2}(5\omega\cos(5\omega) - \sin(5\omega))\sin(\omega ct) e^{i\omega x}\, d\omega.$$

Using Euler's formula, we can write this solution as

$$u(x,t) = -\frac{1}{\pi}\int_{-\infty}^{\infty} \frac{1}{c\omega^3}(5\omega\cos(5\omega) - \sin(5\omega))\sin(\omega ct)\sin(\omega x)\,d\omega.$$

7.3 The Telegraph Equation

2. Since $b^2 - a = -2 < 0$, case 1 applies. Further, $g(x) = 0$, so $F(\omega) = 0$. Compute

$$\begin{aligned}
F(\omega) &= \mathcal{F}[f(x)](\omega) \\
&= \int_{-\infty}^{\infty} f(\xi)e^{-i\omega\xi}\,d\xi \\
&= \int_0^1 (1-\xi)e^{-i\omega\xi}\,d\xi + \int_1^{\infty} (1+\xi)e^{-i\omega\xi}\,d\xi \\
&= -\frac{i\omega - 1 - e^{-i\omega}}{\omega^2} + \frac{e^{-i\omega} + i\omega + 1}{\omega^2} \\
&= \frac{2}{\omega^2}\left(1 - \frac{e^{i\omega} + e^{-i\omega}}{2}\right) \\
&= \frac{2}{\omega^2}(1 - \cos(\omega)).
\end{aligned}$$

Therefore

$$c_1(\omega) = \frac{2}{\omega^2}(1 - \cos(\omega))$$

and

$$c_2(\omega) = \frac{2F(\omega)}{\sqrt{2 + 25\omega^2}} = \frac{4(1 - \cos(\omega))}{\omega^2\sqrt{2 + 25\omega^2}}.$$

Then

$$\begin{aligned}
U(\omega,t) &= c_1(\omega)e^{-2t}\cos\left(\sqrt{2 + 25\omega^2}t\right) \\
&\quad + c_2(\omega)e^{-2t}\sin\left(\sqrt{2 + 25\omega^2}t\right).
\end{aligned}$$

The solution is

$$u(x,t) = \frac{1}{2\pi}\int_{-\infty}^{\infty} U(\omega,t)e^{i\omega x}\,d\omega.$$

4. Here $b^2 - a = 0$, so case 2 applies. Because $f(x) = 0$, $F(\omega) = 0$, so $c_1(\omega) = 0$. Compute

$$\widehat{g}(\omega) = \frac{4}{\omega^2}(\sin(\omega) - \omega\cos(\omega)).$$

Then

$$U(\omega,t) = \frac{2}{\omega^4}(\sin(\omega) - \omega\cos(\omega))e^{-2t}\sin(2\omega t).$$

The solution is

$$u(x,t)=\frac{1}{2\pi}\int_{-\infty}^{\infty}U(\omega,t)e^{i\omega x}\,d\omega.$$

5. Because $b^2-a=13>0$, we are in case 3. Here $F(\omega)=0$ and we find that

$$G(\omega)=\frac{4}{\omega^2}(\sin(\omega)-\omega\cos(\omega)).$$

Let

$$c_1(\omega)=\frac{G(\omega)}{2\sqrt{13-\omega^2}}$$

and $c_2=-c_1$. Let

$$U_1(\omega,t)=e^{-4t}\left[c_1e^{\sqrt{13-\omega^2}t}+c_2e^{-\sqrt{13-\omega^2}t}\right].$$

Let $c_3=0$ and

$$c_4(\omega)=\frac{G(\omega)}{\sqrt{\omega^2-13}}.$$

Let

$$U_2(\omega,t)=e^{-4t}c_4(\omega)\sin(\sqrt{\omega^2-13}t).$$

The solution is

$$u(x,t)=\frac{1}{2\pi}\left[\int_{|\omega|<\sqrt{13}}U_1(\omega,t)e^{i\omega x}\,d\omega+\int_{|\omega|\geq\sqrt{13}}U_2(\omega,t)e^{i\omega x}\,d\omega\right].$$

6. From the solution to problem 4 we know $G(\omega)$. Because we are in case 1, we obtain

$$U(\omega,t)=e^{-2t}c_2(\omega)\sin\left(\sqrt{4+9\omega^2}t\right),$$

where

$$c_2(\omega)=\frac{G(\omega)}{\sqrt{4+9\omega^2}}.$$

The solution is

$$u(x,t)=\frac{1}{2\pi}\int_{-\infty}^{\infty}U(\omega,t)e^{i\omega x}\,d\omega.$$

9. Let $v(x,t)=e^{bt}u(x,t)$ to obtain the problem:

$$\begin{aligned}v_{tt}+(a-b^2)v&=c^2v_{xx}\text{ for }x>0,t<0,\\&=v(x,0)=f(x),v(0,t)=0.\end{aligned}$$

To solve this telegraph equation on the half line, use the Fourier sine transform with respect to x to obtain:

$$\frac{d^2}{dt^2}\widehat{v}_S(\omega,t)+(a-b^2+c^2\omega^2)\widehat{v}_S(\omega,t)=0.$$

To solve this ordinary differential equation in t, look for solutions of the form e^{rt}. The characteristic equation for r is

$$r^2 + (a - b^2 + c^2\omega^2) = 0,$$

with roots

$$r = \pm\sqrt{b^2 - a - c^2\omega^2}.$$

At this point the solution follows the discussion of the telegraph equation on the real line, with three cases for the values of r, and with the sine transform replacing the transform.

7.4 The Laplace Transform

2. Follow the discussion of Section 7.4.2, using the Laplace transform with respect to t, to obtain the solution

$$u(x,t) = \int_0^t (t-\tau)^2 \frac{x}{\sqrt{\pi k}\tau^{3/2}} e^{-x^2/4k\tau}\, d\tau.$$

4. The problem to solve is

$$u_t = ku_{xx} \text{ for } 0 < x < L, t > 0,$$
$$u(x,0) = 1, u(0,t) = u(L,t) = 0.$$

Take the Laplace transform of the differential equation with respect to t and use the operational formula to obtain

$$U''(x,s) - \frac{s}{k}U(x,s) = -\frac{1}{k}.$$

Here primes denote differentiation with respect to x. This ordinary differential equation has general solution

$$U(x,s) = c_1 e^{\sqrt{s/k}x} + c_2 e^{-\sqrt{s/k}x} + \frac{1}{s}.$$

Take the transform of the conditions $u(0,t) = u(L,t) = 0$ to obtain

$$c_1 + c_2 = -\frac{1}{s},$$
$$c_1 e^{\sqrt{s/k}L} + c_2 e^{-\sqrt{s/k}L} = -\frac{1}{s}.$$

Solve these algebraic equations to obtain

$$c_1 = \frac{1}{2s\sinh(\sqrt{s/k}L)}\left(e^{-\sqrt{s/k}L} - 1\right)$$

and

$$c_2 = \frac{1}{2s\sinh(\sqrt{s/k}L)}\left(1 - e^{\sqrt{s/k}L}\right).$$

Substitute these into the expression for $U(x, s)$ and do a fair amount of manipulation to obtain

$$U(x, s) = \frac{-1}{\sinh(\sqrt{s/k}L)}\left[\sinh(\sqrt{s/k}x) - \sinh(\sqrt{s/k}(L - x))\right].$$

Formally, the solution is

$$u(x, t) = \mathcal{L}^{-1}[U(x, s)](x, t).$$

Chapter 8

First-Order Equations

8.1 Linear First-Order Equations

2. The characteristic equation is $dy/dx = -1$, so the characteristics are straight lines $y = -x + k$. Use the transformation

$$\xi = x,\ \eta = x + y.$$

The transformed differential equation is

$$w_\xi + (\eta - \xi)w = 0.$$

Multiply this equation by $e^{\eta\xi - \xi^2/2}$ to write it as

$$\frac{\partial}{\partial \xi}\left(e^{\eta\xi - \xi^2/2}w\right) = 0.$$

Then

$$we^{\eta\xi - \xi^2/2} = g(\eta),$$

and then

$$w(\xi, \eta) = g(\eta)e^{\xi^2/2}e^{-\eta\xi}.$$

In terms of x and y,

$$u(x, y) = g(x + y)e^{-xy - x^2/2},$$

with g any differentiable function.

4. The characteristics are lines $x + 2y = k$. Let

$$\xi = x,\ \eta = x + 2y.$$

Solutions Manual to Accompany Beginning Partial Differential Equations, Third Edition. Peter V. O'Neil.

The transformed equation is

$$w_\xi + \frac{1}{4}(\eta - \xi)w = 0.$$

This has the general solution

$$w(\xi, \eta) = g(\eta)e^{(-\eta\xi/4)+\xi^2/8}.$$

In terms of x and y, the solution is

$$u(x, y) = g(x + 2y)e^{-x^2/8}e^{-xy/2}.$$

6. The characteristic equation is $dy/dx = -2/x^2$ so the characteristics are hyperbolas $y - 2/x = k$. Use the transformation

$$\xi = x,\ \eta = \frac{2}{x} - y.$$

The transformed differential equation is

$$w_\xi - \frac{1}{\xi}w = 1.$$

This has the solution

$$w(\xi, \eta) = \xi \ln(\xi) + \xi g(\eta).$$

Then

$$u(x, y) = x \ln(x) + xg(-y + 2/x)$$

for $x > 0$.

8. The characteristics are lines $y = kx$ through the origin. Let

$$\xi = x,\ \eta = \frac{y}{x}$$

to get

$$\xi w_\xi + w = 1 - \eta.$$

This has the solution

$$w(\xi, \eta) = 1 - \eta + \frac{1}{\xi}g(\eta),$$

so

$$u(x, y) = 1 - \frac{y}{x} + \frac{1}{x}g(y/x).$$

10. The characteristic equation is $dy/dx = -y^2$, so the characteristics are curves $x - 1/y = c$. Use the transformation

$$\xi = x,\ \eta = x - \frac{1}{y}$$

to obtain

$$w_\xi - \frac{1}{\xi - \eta} w = 0.$$

This has the general solution

$$w(\xi, \eta) = (\xi - \eta) g(\eta).$$

Then

$$u(x, y) = \frac{1}{y} g\left(x - \frac{1}{y}\right).$$

12. The characteristic equation is $dy/dx = y/x$, with solutions $y/x = k$. Let

$$\xi = x,\ \eta = \frac{y}{x}.$$

The differential equation transforms to

$$w_\xi = -\frac{2}{\xi}$$

with solution

$$w(\xi, \eta) = -2\ln(\xi) + g(\eta).$$

Then

$$u(x, y) = -2\ln(x) + g\left(\frac{y}{x}\right).$$

8.2 The Significance of Characteristics

2. The characteristics are the lines $y + 6x = k$. Let

$$\xi = x,\ \eta = y + 6x$$

to obtain

$$w_\xi = \eta - 6\xi.$$

Then

$$x(\xi, \eta) = \eta\xi - 3\xi^2 + g(\eta),$$

so

$$u(x, y) = 3x^2 + xy + g(y + 6x).$$

(a) We want a solution satisfying $u(x, y) = e^x$ on the line $y = -6x + 2$. This requires that

$$u(x, 2 - 6x) = 3x^2 + x(2 - 6x) + g(2) = e^x$$

and therefore that

$$g(2) = e^x + 3x^2 - 2x,$$

which is impossible. There is no solution that has the given data on the characteristic $y + 6x = 2$.

(b) We want $u(x, y) = 1$ on $y = -x^2$. Now we need to choose g so that

$$u(x, -x^2) = 1 = 3x^2 + x(-x^2) + g(-x^2 + 6x).$$

This requires that

$$3x^2 - x^3 + g(-x^2 + 6x) = 1.$$

To see how to choose g, let $t = -x^2 + 6x$. Then

$$x = 3 \pm \sqrt{9 - t}.$$

If we use $t = 3 + \sqrt{9 - t}$, then

$$g(t) = 1 + (3 + \sqrt{9-t})^3 - 3(3 + \sqrt{9-t})^2.$$

This choice of g gives us

$$u(x, y) = 3x^2 + xy + 1 + \left(3 + \sqrt{9 - y - 6x}\right)^3 - 3\left(3 + \sqrt{9 - y - 6x}\right)^2.$$

This satisfies $u(x, -x^2) = 1$ if $x > 3$.

If $x < 3$, use $t = 3 - \sqrt{9 - t}$ to obtain

$$u(x, y) = 3x^2 + xy + 1 + \left(3 - \sqrt{9 - y - 6x}\right)^3 = 3\left(3 - \sqrt{9 - y - 6x}\right)^2.$$

(c) We want $u(x, -6x) = -4x$. Now choose g so that

$$u(x, -6x) = -4x = 3x^2 + x(-6x) + g(0).$$

This would have $g(0) = 3x^2 - 4x$, which is impossible. This problem also has no solution along this characteristic.

4. The characteristics are graphs of $x + 2y^2 = k$. Let

$$\xi = x, \ \eta = x + 2y^2$$

to obtain

$$w_\xi + \frac{1}{4}w = 0,$$

with solutions

$$w(\xi, \eta) = g(\eta)e^{-\xi/4}.$$

Then

$$u(x, y) = g(x + 2y^2)e^{-x/4}.$$

(a) We want $u(x, y) = x^3$ on $x + 2y = 4$. We need

$$u(x, (3 - x)/2) = g\left(x + \frac{1}{2}(3 - x)^2\right) e^{-x^2/4} = x^3.$$

Then

$$g\left(\frac{1}{2}x^2 + \frac{9}{2} - 2x\right) = x^3 e^{x/4}.$$

Let $t = x^2/2 + 9/2 - 2x$, so

$$x = 2 \pm \sqrt{2t - 5}.$$

With $x = 2 + \sqrt{2t - 5}$, we get

$$g(t) = (2 + \sqrt{2t - 5})^2 e^{2+\sqrt{2t-5}/4}.$$

Then

$$\begin{aligned} u(x, t) &= g(x + 2y)e^{-x/4} \\ &= \left(2 + \sqrt{2(x + 2y^2) - 5}\right)^3 e^{-x/4} e^{(2+\sqrt{2(x+2y)^2-5})/4}. \end{aligned}$$

This is the solution for $x > 2$. For $x < 2$, use $x = 2 - \sqrt{2t - 5}$ to get

$$u(x, y) = \left(2 - \sqrt{2(x + 2y)^2 - 5}\right)^2 e^{-x/4} e^{(2-\sqrt{2(x+2y^2)-5})/4}.$$

(b) Now we want

$$u(y^2, y) = -y = g(y^2 + 2y^2)e^{-y^2/4}.$$

Then we need

$$g(3y^2) = -ye^{y^2/4}.$$

Let $t = 3y^2$. First using $y = \sqrt{t/3}$, we obtain the solution

$$u(x, y) = -\sqrt{\frac{x + 2y^2}{3}} e^{(x+2y^2)/12} e^{-x/4}.$$

This holds for $y > 0$. If we use $y = -\sqrt{t/3}$, so $y < 0$, we obtain

$$u(x, y) = \sqrt{\frac{x + 2y^2}{3}} e^{(x+2y^2)/12} e^{-x/4}.$$

(c) We want $u(x, y) = 2$ on $x + 2y^2 = 1$. This requires that

$$u(x, y) = 2 - g(1)e^{-x/4},$$

and this is impossible for any g. There is no solution with information specified along this characteristic.

6. The characteristics are curves $x^3 - y^3 = k$. Let

$$\xi = x, \; \eta = x^3 - y^3.$$

The transformed equation is $w_\xi = 1$, so

$$w(\xi, \eta) = \xi + g(\eta),$$

and

$$u(x, y) = x + g(x^3 - y^3).$$

(a) To have $u(x, 4x) = x$, we need

$$u(x, 4x) = x + g(x^3 - 64x^3) = x + g(-63x^3) = x.$$

We can have this by choosing g to be identically zero. The solution is $u(x, y) = x$.

(b) To have $u(x, y) = -2y$ for $y^3 = x^3 - 2$, we need

$$u(x, y) = x + g(2) = -2y.$$

This is impossible for any g. This problem has no solution.

(c) Finally, to have $u(x, -x) = y^2$, we need

$$u(x, -x) = x + g(x^3 - (-x)^3) = x + g(2x^3) = y^2.$$

Then

$$g(2x^3) = y^3 - x = x^2 - x.$$

Let $t = 2x^2$. Then $x = (t/2)^{1/3}$ and

$$g(t) = (t/2)^{2/3} - (t/2)^{1/3}.$$

In this case the solution is

$$u(x, y) = x + \left(\frac{x^2 - y^2}{2}\right)^{2/3} - \left(\frac{x^2 - y^2}{2}\right)^{1/3}.$$

8.3 The Quasi-Linear Equation

2. The characteristics are determined by

$$\frac{dx}{dt} = 1, \; \frac{dy}{dt} = -x, \; \frac{du}{dt} = 4.$$

Integrate these to obtain

$$x = t + A, \; y = -\frac{1}{2}t^2 - At + B, \; u = 4t + C.$$

Suppose that a characteristic intersects Γ at $P : (s, 4s, 0)$ at $t = 0$. Then, at $t = 0$,

$$x = s = A,\ y = 4s = B,\ u = C = 0.$$

Then

$$x = t + s,\ y = -\frac{1}{2}t^2 - st + 4s,\ u = 4t.$$

We must eliminate s and t from these equations. Begin with $t = u/4$ to write

$$s = x - t = x - \frac{u}{4}.$$

Then

$$\begin{aligned} y &= -\frac{1}{2}\frac{u^2}{16} - \left(x - \frac{u}{4}\right)\frac{u}{4} + 4\left(x - \frac{u}{4}\right) \\ &= -\frac{1}{32}u^2 - \frac{1}{4}ux + 4x - u. \end{aligned}$$

This equation in x, y and u implicitly defines the solution $u(x, y)$.

4. The characteristics are found to be

$$x = t + A,\ -\frac{1}{2y^2} = t + B,\ \sin(u) = t + C.$$

If a characteristic intersects Γ at $P : (s, s^2, 0)$ when $t = 0$, we get

$$x = s = A,\ -\frac{1}{2}s^4 = B,\ C = 0.$$

Upon eliminating s and t from these equations, we get

$$-\frac{1}{2y^2} = \sin(u) - \frac{1}{2}(x - \sin(u))^{-4}.$$

This equation implicitly defines $u(x, y)$.

6. The characteristics are given by

$$x = t + A,\ -\frac{1}{y} = t + B,\ \ln|\sec(u) + \tan(u)| = t + C.$$

Suppose a characteristic intersects Γ at $P : (s^2, s, 0)$. Then

$$x = s^2 = A,\ -\frac{1}{y} = -\frac{1}{s} = B,\ \ln(1) = 0 = C.$$

Then

$$x = t + s^2,\ -\frac{1}{y} = t - \frac{1}{s},\ \ln|\sec(u) + \tan(u)| = t.$$

Eliminate s and t to obtain the equation

$$-\frac{1}{y} = \ln|\sec(u) + \tan(u)| - \frac{1}{(x - \ln|\sec(u) + \tan(u)|)^{1/2}}.$$

This implicitly defines $u(x, y)$.

8. The characteristics are determined by

$$-\frac{1}{2x^2} = t + A, \; -\ln(y) = t + B, \; \ln(u) = t + C.$$

Suppose a characteristic intersects Γ at $P : (s^2 - 1, s, 1)$ when $t = 0$. Then

$$-\frac{1}{2(s^2-1)^2} = A, \; -\ln(s) = B, \; C = 0.$$

Then

$$-\frac{1}{2x^2} = t - \frac{1}{2(s^2-1)^2}, \; -\ln(y) = t - \ln(s), \; \ln(u) = t.$$

Eliminate s and t from this equation to obtain

$$-\frac{2}{2x^2} = \ln(u) - \frac{1}{2(y^2u^2-1)^2}.$$

This defines $u(x, y)$.

10. The characteristics are given by

$$x = Ct + A, \; y = t + B, \; u = C.$$

Consider Γ as the locus of points $(s, 0, f(s))$. Suppose that a characteristic intersects Γ at $P : (s, 0, f(s))$ when $t = 0$. Then

$$u = C = f(s), \; x - s = A, \; y = B = 0.$$

Then

$$u = f(s), \; x = tf(s) = x - yu.$$

Then

$$u(x, y) = f(x - yu(x, y)).$$

This implicitly defines the solution.

12. (b) With $f(u) = u$, the solution is defined implicitly by

$$u(x, y) = \varphi(x + uy).$$

(d) With $f(u) = e^u$, the solution is implicitly defined by

$$u(x, y) = \varphi(x + e^u y).$$

PURE AND APPLIED MATHEMATICS

A Wiley Series of Texts, Monographs, and Tracts

Founded by RICHARD COURANT

Editors Emeriti: MYRON B. ALLEN III, PETER HILTON, HARRY HOCHSTADT, ERWIN KREYSZIG, PETER LAX, JOHN TOLAND

ADÁMEK, HERRLICH, and STRECKER—Abstract and Concrete Catetories
ADAMOWICZ and ZBIERSKI—Logic of Mathematics
AINSWORTH and ODEN—A Posteriori Error Estimation in Finite Element Analysis
AKIVIS and GOLDBERG—Conformal Differential Geometry and Its Generalizations
ALLEN and ISAACSON—Numerical Analysis for Applied Science
*ARTIN—Geometric Algebra
ATKINSON, HAN, and STEWART—Numerical Solution of Ordinary Differential Equations
AUBIN—Applied Functional Analysis, Second Edition
AZIZOV and IOKHVIDOV—Linear Operators in Spaces with an Indefinite Metric
BASENER—Topology and Its Applications
BERG—The Fourier-Analytic Proof of Quadratic Reciprocity
BERKOVITZ—Convexity and Optimization in $\mathbb{R}^n$
BERMAN, NEUMANN, and STERN—Nonnegative Matrices in Dynamic Systems
BOYARINTSEV—Methods of Solving Singular Systems of Ordinary Differential Equations
BRIDGER—Real Analysis: A Constructive Approach
BURK—Lebesgue Measure and Integration: An Introduction
*CARTER—Finite Groups of Lie Type
CASTILLO, COBO, JUBETE, and PRUNEDA—Orthogonal Sets and Polar Methods in Linear Algebra: Applications to Matrix Calculations, Systems of Equations, Inequalities, and Linear Programming
CASTILLO, CONEJO, PEDREGAL, GARCIÁ, and ALGUACIL—Building and Solving Mathematical Programming Models in Engineering and Science
CHATELIN—Eigenvalues of Matrices
CLARK—Mathematical Bioeconomics: The Mathematics of Conservation, Third Edition
COX—Galois Theory, Second Edition
COX—Primes of the Form $x^2 + ny^2$: Fermat, Class Field Theory, and Complex Multiplication, Second Edition
*CURTIS and REINER—Representation Theory of Finite Groups and Associative Algebras
*CURTIS and REINER—Methods of Representation Theory: With Applications to Finite
Groups and Orders, Volume I
CURTIS and REINER—Methods of Representation Theory: With Applications to Finite Groups and Orders, Volume II
DINCULEANU—Vector Integration and Stochastic Integration in Banach Spaces
*DUNFORD and SCHWARTZ—Linear Operators
Part 1—General Theory
Part 2—Spectral Theory, Self Adjoint Operators in Hilbert Space
Part 3—Spectral Operators

*Now available in a lower priced paperback edition in the Wiley Classics Library.
†Now available in paperback.

FARINA and RINALDI—Positive Linear Systems: Theory and Applications
FATICONI—The Mathematics of Infinity: A Guide to Great Ideas, Second Edition
FOLLAND—Real Analysis: Modern Techniques and Their Applications
FRÖLICHER and KRIEGL—Linear Spaces and Differentiation Theory
GARDINER—Teichmüller Theory and Quadratic Differentials
GILBERT and NICHOLSON—Modern Algebra with Applications, Second Edition
*GRIFFITHS and HARRIS—Principles of Algebraic Geometry
GRILLET—Algebra
GROVE—Groups and Characters
GUSTAFSSON, KREISS, and OLIGER—Time-Dependent Problems and Difference Methods, Second Edition
HANNA and ROWLAND—Fourier Series, Transforms, and Boundary Value Problems, Second Edition
*HENRICI—Applied and Computational Complex Analysis
Volume 1, Power Series—Integration—Conformal Mapping—Location of Zeros
Volume 2, Special Functions—Integral Transforms—Asymptotics—Continued Fractions
Volume 3, Discrete Fourier Analysis, Cauchy Integrals, Construction of Conformal Maps, Univalent Functions
*HILTON and WU—A Course in Modern Algebra
*HOCHSTADT—Integral Equations
JOST—Two-Dimensional Geometric Variational Procedures
KHAMSI and KIRK—An Introduction to Metric Spaces and Fixed Point Theory
*KOBAYASHI and NOMIZU—Foundations of Differential Geometry, Volume I
*KOBAYASHI and NOMIZU—Foundations of Differential Geometry, Volume II
KOSHY—Fibonacci and Lucas Numbers with Applications
LAX—Functional Analysis
LAX—Linear Algebra and Its Applications, Second Edition
LOGAN—An Introduction to Nonlinear Partial Differential Equations, Second Edition
LOGAN and WOLESENSKY—Mathematical Methods in Biology
LUI—Numerical Analysis of Partial Differential Equations
MARKLEY—Principles of Differential Equations
MORRISON—Functional Analysis: An Introduction to Banach Space Theory
NAYFEH—Perturbation Methods
NAYFEH and MOOK—Nonlinear Oscillations
O'LEARY—Revolutions of Geometry
O'NEIL—Beginning Partial Differential Equations, Third Edition
O'NEIL—Solutions Manual to Accompany Beginning Partial Differential Equations, Third Edition
PANDEY—The Hilbert Transform of Schwartz Distributions and Applications
PETKOV—Geometry of Reflecting Rays and Inverse Spectral Problems
*PRENTER—Splines and Variational Methods
PROMISLOW—A First Course in Functional Analysis
RAO—Measure Theory and Integration
RASSIAS and SIMSA—Finite Sums Decompositions in Mathematical Analysis
RENELT—Elliptic Systems and Quasiconformal Mappings
RIVLIN—Chebyshev Polynomials: From Approximation Theory to Algebra and Number Theory, Second Edition
ROCKAFELLAR—Network Flows and Monotropic Optimization
ROITMAN—Introduction to Modern Set Theory

*Now available in a lower priced paperback edition in the Wiley Classics Library.
†Now available in paperback.

ROSSI—Theorems, Corollaries, Lemmas, and Methods of Proof
*RUDIN—Fourier Analysis on Groups
SENDOV—The Averaged Moduli of Smoothness: Applications in Numerical Methods and Approximations
SENDOV and POPOV—The Averaged Moduli of Smoothness
SEWELL—The Numerical Solution of Ordinary and Partial Differential Equations, Second Edition
SEWELL—Computational Methods of Linear Algebra, Second Edition
SHICK—Topology: Point-Set and Geometric
SHISKOWSKI and FRINKLE—Principles of Linear Algebra With *Maple*™
SHISKOWSKI and FRINKLE—Principles of Linear Algebra With *Mathematica*®
*SIEGEL—Topics in Complex Function Theory
Volume 1—Elliptic Functions and Uniformization Theory
Volume 2—Automorphic Functions and Abelian Integrals
Volume 3—Abelian Functions and Modular Functions of Several Variables
SMITH and ROMANOWSKA—Post-Modern Algebra
ŠOLÍN–Partial Differential Equations and the Finite Element Method
STADE—Fourier Analysis
STAHL and STENSON—Introduction to Topology and Geometry, Second Edition
STAHL—Real Analysis, Second Edition
STAKGOLD and HOLST—Green's Functions and Boundary Value Problems, Third Edition
STANOYEVITCH—Introduction to Numerical Ordinary and Partial Differential Equations Using MATLAB®
*STOKER—Differential Geometry
*STOKER—Nonlinear Vibrations in Mechanical and Electrical Systems
*STOKER—Water Waves: The Mathematical Theory with Applications
WATKINS—Fundamentals of Matrix Computations, Third Edition
WESSELING—An Introduction to Multigrid Methods
†WHITHAM—Linear and Nonlinear Waves
ZAUDERER—Partial Differential Equations of Applied Mathematics, Third Edition

*Now available in a lower priced paperback edition in the Wiley Classics Library.
†Now available in paperback.